Cram101 Textbook Outlines to accompany:

# Theory and Practice of Water and Wastewater Treatment

## Droste, 1st Edition

A Content Technologies Inc. publication (c) 2012.

## Learning System

Cram101 Textbook Outlines is a learning system. The notes in this book are the highlights of your textbook, you will never have to highlight a book again.

**How to use this book.** Take this book to class, it is your notebook for the lecture. The notes and highlights on the left hand side of the pages follow the outline and order of the textbook. All you have to do is follow along while your instructor presents the lecture. Circle the items emphasized in class and add other important information on the right side. With Cram101 Textbook Outlines you'll spend less time writing and more time listening. Learning becomes more efficient.

## Cram101.com Online

Increase your studying efficiency by using Cram101.com's practice tests and online reference material. It is the perfect complement to Cram101 Textbook Outlines. Use self-teaching matching tests or simulate in-class testing with comprehensive multiple choice tests, or simply use Cram's true and false tests for quick review. Cram101.com even allows you to enter your in-class notes for an integrated studying format combining the textbook notes with your class notes.

Visit **www.Cram101.com**, click Sign Up at the top of the screen, and enter **DK73DW1208** in the promo code box on the registration screen. Your access to www.Cram101.com is discounted by 50% because you have purchased this book. Sign up and stop highlighting textbooks forever.

Theory and Practice of Water and Wastewater Treatment
Droste, 1st

# CONTENTS

## Chapter 1. Basic Chemistry

| | |
|---|---|
| Element1 | A pure substance that cannot be split up into anything simpler is called an element. |
| Atom1 | Atom refers to the smallest particle of an element which can exist. |
| Periodic1 | A dynamical system or function which at some point returns to the same state or value. If a system or function ever revisits the identical state or value it will continue to come back to it again and again in equal intervals of time. That is why we call such a system periodic. |
| Atomic number1 | Number of protons in the nucleus of the atom is an atomic number. |
| Atomic weight1 | Approximately the sum of the number of protons and neutrons found in the nucleus of an atom is the atomic weight. |
| Weight1 | Force of gravity of an object is called weight. |
| Isotopes1 | Atoms with the same atomic number, but different mass numbers are called isotopes. |
| Isotope1 | Atomic nuclei having same number of protons but different numbers of neutrons are called an isotope. |
| Precision1 | Degree of exactness in a measurement is precision. |
| Quantity1 | A numerical value either scalar or vector, which describes some attribute of an object like its position or its velocity . We sometimes speak of physical quantities to signify that we are talking about an object's properties or attributes as opposed to a purely mathematical quantity. |
| Mole1 | The amount of substance containing $6 \times 10^{23}$ particles is the mole. |
| Liter1 | A metric system unit of volume, usually used for liquids is referred to as liter. |
| Density1 | The mass of a given volume of substance. It has units of kg/m3 or g/cm3. When the density is high the particles are closely packed. |
| Temperature1 | Temperature refers to measure of hotness of object on a quantitative scale. In gases, proportional to average kinetic energy of molecules. |

## Chapter 1. Basic Chemistry

| | |
|---|---|
| Equation1 | An equation is a mathematical expression with an equal sign in it. It signifies that the numerical or vector value on one side of the = is the same as the numerical or vector value on the other side. An equation may include variables and parameters. If any of the variables are rates of change, the equation is called a differential equation. |
| Charge1 | A property of atomic particles. Electrons and protons have opposite charges and attract each other. The charge on an electron is negative and that on a proton is positive. Measured in coulombs. |
| Charged1 | Object that has an unbalance of positive and negative electrical charges is referred to as charged. |
| Ion1 | Ion refers to a positively or negatively charged particle formed when an atom or group of atoms loses or gains electrons. |
| Solid1 | State of matter with fixed volume and shape is referred to as solid. |
| Phase1 | The particles in a wave, which are in the same state of vibration, i.e., the same position and the same direction of motion are said to be in the same phase. |
| Covalent bond1 | A chemical bond formed by the sharing of a pair of electrons is called a covalent bond. |
| Attract1 | To pull together is to attract. |
| Negative1 | The sign of the electric charge on the electron is negative. |
| Periodic table1 | Periodic table refers to a chart of all the known chemical elements arranged according to the number of protons in the nucleus . Elements with similar properties are grouped together in the same column. |
| Secondary1 | The output coil of a transformer is referred to as secondary. |
| Viscosity1 | A measure of how easy or otherwise a liquid can be poured. A high viscosity liquid flows very slowly. |
| Melting point1 | Melting point refers to temperature at which substance changes from solid to liquid state. |
| Resonance1 | When the frequency of an external force matches the natural frequency and standing waves are set up, we have resonance. |

## Chapter 1. Basic Chemistry

| | |
|---|---|
| Stable1 | Does not decay is stable. |
| Energy1 | Energy refers to non-material property capable of causing changes in matter. |
| Proton1 | Subatomic particle with positive charge that is nucleus of hydrogen atom is called proton. |
| Properties1 | Properties refers to qualities or attributes that, taken together, are usually unique to an object; for example, color, texture, and size. |
| Molecule1 | The smallest part of an element or compound that can exist on its own is called the molecule. |
| Boiling point1 | Temperature at which a substance, under normal atmospheric pressure, changes from a liquid to a vapor state is the boiling point. |
| Heat of vaporization1 | Heat of vaporization refers to quantity of energy needed to change a unit mass of a substance from liquid to gaseous state at the boiling point. |
| Heat of fusion1 | Heat of fusion refers to quantity of energy needed to change a unit mass of a substance from solid to liquid state at the melting point. |
| Fusion1 | Fusion refers to combination of two nuclei into one with release of energy. |
| Liquids1 | Liquids refers to a phase of matter composed of molecules that have interactions stronger than those found in a gas but not strong enough to keep the molecules near the equilibrium positions of a solid, resulting in the characteristic definite volume but indefinite shape. |
| Electronegativity1 | The comparative ability of atoms of an element to attract bonding electrons is referred to as electronegativity. |
| Neutral1 | Object that has no net electric charge is called neutral. |
| State1 | Dynamical systems evolve over the course of time. The state of the system at any instant may be identified by the values of certain variables at that instant. For example specifying the angle from the vertical and the velocity of a frictionless pendulum allows us to predict its position and velocity at any future time. Therefore the state of the pendulum at any instant is its position and velocity. In this example the position and velocity are known as state variables. |
| Second1 | Second is a SI unit of time. |

## Chapter 1. Basic Chemistry

| | |
|---|---|
| Mercury1 | Mercury refers to the innermost planet in the solar system, and a metallic element that is liquid at room temperature. |
| Force1 | Force refers to agent that results in accelerating or deforming an object. |
| Electron1 | Subatomic particle of small mass and negative charge found in every atom is called the electron. |
| Matter1 | We call the commonly observed particles such as protons, neutrons and electrons matter particles, and their antiparticles, antimatter. |
| Core1 | In electromagnetism, the material inside the coils of a transformer or electromagnet is a core. |
| Equilibrium1 | Equilibrium refers to condition in which net force is equal to zero. Condition in which net torque on object is zero. |
| Function1 | A mathematical function is a rule relating two sets of objects. Here we will restrict ourselves to objects that are numbers or vectors. One of the sets is called the domain of the function, the other is called the range of the function. Functions are frequently expressed as equations as for example $Y=X+2$. This function is interpreted as follows. For every X in the domain, add 2 to it to get the corresponding Y in the range. Because we are free to choose any X we want, X is called the independent variable. Because once X is chosen Y is fixed, we call Y the dependent variable. |
| Units1 | The units one uses should be of a size that makes sense for the particular subject at hand. It is easiest to define units in each area of science and then relate them to one another than to go around measuring particle masses in grams or cheese in proton mass units. In particle physics the standard unit is the unit of energy gev. One ev is the amount of energy that an electron gains when it moves through a potential difference of 1 Volt . G stands for Giga, or $10^9$. Thus a gev is a billion electron Volts. The mass-energy of a proton or neutron is approximately 1 gev. |
| Gases1 | A phase of matter composed of molecules that are relatively far apart moving freely in a constant, random motion and have weak cohesive forces acting between them, resulting in the characteristic indefinite shape and indefinite volume of a gas are called the gases. |
| Liquid1 | Material that has fixed volume but whose shape depends on the container is called liquid. |
| Current1 | Current refers to a flow of charge. Measured in amps. |

CTam101

## Chapter 1. Basic Chemistry

| | |
|---|---|
| Voltage drop1 | The electric potential difference across a resistor or other part of a circuit that consumes power is referred to as voltage drop. |
| Ohm's law1 | Ohm's law refers to the ratio of the potential difference between the ends of a conductor to the current flowing through it is constant; the constant of proportionality is called the resistance, and is different for different materials. |
| Resistance1 | Ratio of potential difference across device to current through it are called the resistance. |
| Distance1 | Distance refers to separation between two points. A scalar quantity. |
| Ohm1 | Ohm refers to SI unit of resistance; one volt per ampere. |
| Normal1 | Perpendicular to plane of interest is called normal. |
| Interaction1 | A process in which a particle decays or it responds to a force due to the presence of another particle is called interaction. |
| Solids1 | A phase of matter with molecules that remain close to fixed equilibrium positions due to strong interactions between the molecules, resulting in the characteristic definite shape and definite volume are solids. |
| Activity1 | Number of decays per second of a radioactive substance is an activity. |
| Velocity1 | The ratio of change in position with respect to the time interval over which the change occurred is referred to as velocity. |
| Decay1 | Any process in which a particle disappears and in its place two or more different particles appear is decay. |
| Speed1 | Speed refers to ratio of distance traveled to time interval. |
| Energy levels1 | Amounts of energy an electron in an atom may have are called the energy levels. |
| Frequency1 | Frequency refers to number of occurrences per unit time. |
| Gas1 | State of matter that expands to fill container is referred to as gas. |

## Chapter 1. Basic Chemistry

| | |
|---|---|
| Boyle's law1 | The product of the pressure and the volume of an ideal gas at constant temperature is a constant is Boyle's law. |
| Pressure1 | Force per unit area is referred to as the pressure. |
| Mixture1 | Matter made of unlike parts that have a variable composition and can be separated into their component parts by physical means is a mixture. |
| Ideal gas equation1 | An equation which sums up the ideal gas laws in one simple equation, $P V = n R T$, where P is the pressure, V is the volume, n is the number of moles present, and T is the temperature of the sample is called ideal gas equation. |
| Supersaturated1 | Containing more than the normal saturation amount of a solute at a given temperature is supersaturated. |
| Metal1 | Matter having the physical properties of conductivity, malleability, ductility, and luster is referred to as metal. |
| Set1 | In mathematics a set is a collection of related objects. The mathematical usage is similar to the ordinary English meaning of the word. The objects that make up a set are called the elements of the set. If a set contains an unlimited number of elements it is an infinite set. Otherwise it is a finite set. |
| Neutron1 | Neutron refers to subatomic particle with no charge and mass slightly greater than that of proton; type of nucleon. |
| Unstable1 | Matter that is capable of undergoing spontaneous change, as in a radioactive nuclide or an excited nuclear system. An unstable particle is any elementary particle that spontaneously decays into other particles. |
| Radiation1 | Electromagnetic waves that carry energy are referred to as radiation. |
| Alpha decay1 | Process in which a nucleus emits an alpha particle is called alpha decay. |
| Beta decay1 | Beta decay refers to radioactive decay process in which an electron or positron and neutrino is emitted from a nucleus. |
| Nucleus1 | Nucleus refers to the central part of an atom. It contains all the positive charge and almost all the mass of the atom. |

## Chapter 1. Basic Chemistry

| | |
|---|---|
| Power1 | Power refers to rate of doing work; rate of energy conversion. |
| Positron1 | Antiparticle equivalent of electron is a positron. |
| Particle1 | In 'particle physics', a subatomic object with definite mass and charge . |
| Gamma ray1 | Very short wavelength electromagnetic radiation given off from an atomic nucleus is referred to as a gamma ray. |
| Electromagnetic radiation1 | Energy carried by electromagnetic waves throughout space is electromagnetic radiation. |
| Photon1 | Photon refers to quantum of electromagnetic waves; particle aspect of these waves. |
| Speed of light1 | Speed of light in vacuum, $2.9979458 * 10^8$ m/s. |
| Light1 | Electromagnetic radiation with wavelengths between 400 and 700 nm that is visible is called light. |
| Radioactive decay1 | Spontaneous change of unstable nuclei into other nuclei is radioactive decay. |
| Uranium1 | A mildly radioactive element with two isotopes which are fissile and two, which are fertile . Uranium is the basic raw material of nuclear energy. |
| Radioactivity1 | The property of spontaneously emitting alpha, beta, and/or gamma radiation as a result of nuclear disintegration is radioactivity. |
| Nuclide1 | Nucleus of an isotope is referred to as nuclide. |
| Radium1 | An element often found in uranium ore. It has several radioactive isotopes. Radium-226 decays to Radon-222. |
| Radon1 | A heavy radioactive gas given off by rocks containing radium is called radon. |
| Daughter1 | A nucleus formed by the radioactive decay of a different nuclide is called the daughter. |
| Curie1 | The basic unit used to describe the intensity of radioactivity in a sample of material. One curie equals thirty-seven billion disintegrations per second, or approximately the radioactivity of one gram of radium. |

## Chapter 1. Basic Chemistry

| Emitter1 | A dark-coloured object is good emitter of infrared radiation. Light-coloured and silvered objects are poor emitters. |
| --- | --- |
| Photons1 | Photons refers to a quanta of energy in light wave; the particle associated with light . |
| Becquerel1 | The SI unit of radioactivity equal one disintegration per second and symbolized as Bq, is the becquerel. |
| Nuclear transformation1 | Nuclear transformation refers to the process by which an atomic nucleus is transformed into another type of atomic nucleus. For example, by removing an alpha particle from the nucleus, the element radium is transformed into the element radon. |
| System1 | Defined collection of objects is called a system. |
| Physics1 | Study of matter and energy and their relationship is physics. |
| Groundwater1 | Groundwater refers to water found in the voids or free space of soils and rocks underground. |

## Chapter 2. The Thermodynamic Basis for Equilibrium

| | |
|---|---|
| Equilibrium1 | Equilibrium refers to condition in which net force is equal to zero. Condition in which net torque on object is zero. |
| Force1 | Force refers to agent that results in accelerating or deforming an object. |
| State1 | Dynamical systems evolve over the course of time. The state of the system at any instant may be identified by the values of certain variables at that instant. For example specifying the angle from the vertical and the velocity of a frictionless pendulum allows us to predict its position and velocity at any future time. Therefore the state of the pendulum at any instant is its position and velocity. In this example the position and velocity are known as state variables. |
| Energy1 | Energy refers to non-material property capable of causing changes in matter. |
| System1 | Defined collection of objects is called a system. |
| Second1 | Second is a SI unit of time. |
| First law of thermodynamics1 | First law of thermodynamics refers to change in internal or thermal energy is equal to heat added and work done on system. Same as law of conservation of energy. |
| Heat1 | Heat refers to quantity of energy transferred from one object to another because of a difference in temperature. |
| Kinetic energy1 | Energy of object due to its motion is kinetic energy. |
| Friction1 | Force opposing relative motion of two objects are in contact is friction. |
| Convection1 | Heat transfer by means of motion of fluid is a convection. |
| Sun1 | Sun refers to the star at the centre of the solar system. The star the Earth orbits. |
| Waste1 | High-level waste is highly radioactive material arising from nuclear fission. It is recovered from reprocessing spent fuel, though some countries regard spent fuel itself as HLW and plan to dispose of it in that form. It requires very careful handling, storage and disposal. |
| Explosion1 | Explosion refers to a very rapid reaction accompanied by a large expansion of gases. |
| Entropy1 | Entropy refers to a measure of disorder in a system; ratio of heat added to temperature. |

## Chapter 2. The Thermodynamic Basis for Equilibrium

| | |
|---|---|
| Internal energy1 | Sum of all the potential energy and all the kinetic energy of all the molecules of an object is an internal energy. |
| Function1 | A mathematical function is a rule relating two sets of objects. Here we will restrict ourselves to objects that are numbers or vectors. One of the sets is called the domain of the function, the other is called the range of the function. Functions are frequently expressed as equations as for example Y=X+2. This function is interpreted as follows. For every X in the domain, add 2 to it to get the corresponding Y in the range. Because we are free to choose any X we want, X is called the independent variable. Because once X is chosen Y is fixed, we call Y the dependent variable. |
| Gas1 | State of matter that expands to fill container is referred to as gas. |
| Temperature1 | Temperature refers to measure of hotness of object on a quantitative scale. In gases, proportional to average kinetic energy of molecules. |
| Activity1 | Number of decays per second of a radioactive substance is an activity. |
| Pressure1 | Force per unit area is referred to as the pressure. |
| Gases1 | A phase of matter composed of molecules that are relatively far apart moving freely in a constant, random motion and have weak cohesive forces acting between them, resulting in the characteristic indefinite shape and indefinite volume of a gas are called the gases. |
| Equation1 | An equation is a mathematical expression with an equal sign in it. It signifies that the numerical or vector value on one side of the = is the same as the numerical or vector value on the other side. An equation may include variables and parameters. If any of the variables are rates of change, the equation is called a differential equation. |
| Power1 | Power refers to rate of doing work; rate of energy conversion. |
| Liquid1 | Material that has fixed volume but whose shape depends on the container is called liquid. |
| Solid1 | State of matter with fixed volume and shape is referred to as solid. |
| Element1 | A pure substance that cannot be split up into anything simpler is called an element. |
| Electron1 | Subatomic particle of small mass and negative charge found in every atom is called the electron. |

## Chapter 2. The Thermodynamic Basis for Equilibrium

| | |
|---|---|
| Set1 | In mathematics a set is a collection of related objects. The mathematical usage is similar to the ordinary English meaning of the word. The objects that make up a set are called the elements of the set. If a set contains an unlimited number of elements it is an infinite set. Otherwise it is a finite set. |
| Units1 | The units one uses should be of a size that makes sense for the particular subject at hand. It is easiest to define units in each area of science and then relate them to one another than to go around measuring particle masses in grams or cheese in proton mass units. In particle physics the standard unit is the unit of energy gev. One ev is the amount of energy that an electron gains when it moves through a potential difference of 1 Volt . G stands for Giga, or $10^9$. Thus a gev is a billion electron Volts. The mass-energy of a proton or neutron is approximately 1 gev. |
| Phase1 | The particles in a wave, which are in the same state of vibration, i.e., the same position and the same direction of motion are said to be in the same phase. |
| Graphite1 | A form of carbon used in very pure form as a moderator, principally in gas-cooled reactors, but also in Soviet-designed RBMK reactors is graphite. |
| Liter1 | A metric system unit of volume, usually used for liquids is referred to as liter. |
| Mole1 | The amount of substance containing 6 x $10^{23}$ particles is the mole. |
| Solids1 | A phase of matter with molecules that remain close to fixed equilibrium positions due to strong interactions between the molecules, resulting in the characteristic definite shape and definite volume are solids. |
| Density1 | The mass of a given volume of substance. It has units of kg/m3 or g/cm3. When the density is high the particles are closely packed. |
| Crystalline1 | A regularly repeated crystal-like substructure is crystalline. |
| Vapor1 | The gaseous state of a substance that is normally in the liquid state is vapor. |
| Negative1 | The sign of the electric charge on the electron is negative. |
| Matter1 | We call the commonly observed particles such as protons, neutrons and electrons matter particles, and their antiparticles, antimatter. |

## Chapter 2. The Thermodynamic Basis for Equilibrium

| | |
|---|---|
| Ion1 | Ion refers to a positively or negatively charged particle formed when an atom or group of atoms loses or gains electrons. |
| Electrical energy1 | Electrical energy refers to a form of energy from electromagnetic interactions; one of five forms of energy-mechanical, chemical, radiant, electrical, and nuclear . |
| Cell1 | In electricity, a cell is a combination of metals and chemicals that produces a voltage and can cause a current. |
| Period1 | Period refers to time needed to repeat one complete cycle of motion. |
| Barrier1 | Radiation-absorbing material, such as lead or concrete, used to reduce radiation exposure. A primary barrier attenuates useful beam to the required degree. A secondary barrier attenuates stray radiation to the required degree. |
| Charge1 | A property of atomic particles. Electrons and protons have opposite charges and attract each other. The charge on an electron is negative and that on a proton is positive. Measured in coulombs. |
| Current1 | Current refers to a flow of charge. Measured in amps. |
| Potentiometer1 | Potentiometer refers to electrical device with variable resistance; rheostat. |
| Voltage1 | Voltage refers to potential difference. It is a measure of the change in energy that one coulomb of electric charge undergoes when moved between 2 points. |
| Stable1 | Does not decay is stable. |
| Cathode1 | Cathode refers to a negatively charged electrode. |
| Faraday constant1 | Faraday constant refers to the electric charge carried by one mole of electrons . It is equal to the product of the Avogadro constant and the charge on an electron. |
| Potential difference1 | Difference in electric potential between two points is called the potential difference. |
| Voltmeter1 | A device used to measure voltage is called voltmeter. |

## Chapter 2. The Thermodynamic Basis for Equilibrium

| | |
|---|---|
| Range1 | The range is the set of values that the dependent variable of a function may take on. A range may be finite as in the set of numbers {1, 2, 3.n} or infinite as in all the mumbers between 0 and 1. |
| Meter1 | Meter refers to SI unit of length. |
| Metal1 | Matter having the physical properties of conductivity, malleability, ductility, and luster is referred to as metal. |
| Electromotive force1 | Potential difference produced by electromagnetic induction is called electromotive force. |
| Position1 | Separation between object and a reference point is a position. |
| Ionized1 | An atom or a particle that has a net charge because it has gained or lost electrons is ionized. |
| Direct current1 | Current that does not change direction is referred to as direct current. |
| Resistance1 | Ratio of potential difference across device to current through it are called the resistance. |
| Polarization1 | A polarized particle beam is a beam of particles whose spins are aligned in a particular direction. The polarization of the beam is the fraction of the particles with the desired alignment. |
| Grounding1 | Process of connecting a charged object to Earth to remove object's unbalanced charge are referred to as grounding. |
| Insulator1 | Material through which the flow of electrical charge carriers or heat is greatly reduced is referred to as an insulator. |
| Color1 | The visual perception of light that enables human eyes to differentiate between wavelengths of the visible spectrum, with the longest wavelengths appearing red and the shortest appearing blue or violet is called color. |
| Weight1 | Force of gravity of an object is called weight. |
| Proportionality constant1 | A constant applied to a proportionality statement that transforms the statement into an equation is a proportionality constant. |

## Chapter 2. The Thermodynamic Basis for Equilibrium

Efficiency1 | Ratio of output work to input work is called efficiency.

Switch1 | Switch refers to a device in an electrical circuit that breaks or makes a complete path for the current.

Normal1 | Perpendicular to plane of interest is called normal.

Conductor1 | Conductor refers to materials through which charged particles move readily; or heat flow readily.

Kilogram1 | SI unit of mass is referred to as kilogram.

## Chapter 3. Acid - Base Chemistry

| | |
|---|---|
| State1 | Dynamical systems evolve over the course of time. The state of the system at any instant may be identified by the values of certain variables at that instant. For example specifying the angle from the vertical and the velocity of a frictionless pendulum allows us to predict its position and velocity at any future time. Therefore the state of the pendulum at any instant is its position and velocity. In this example the position and velocity are known as state variables. |
| Equilibrium1 | Equilibrium refers to condition in which net force is equal to zero. Condition in which net torque on object is zero. |
| Magnitude1 | Magnitude refers to the size of a thing, without regard for its sign or direction. Similar to the absolute value of a number but applies to vectors as well. |
| Range1 | The range is the set of values that the dependent variable of a function may take on. A range may be finite as in the set of numbers {1, 2, 3.n} or infinite as in all the mumbers between 0 and 1. |
| Ion1 | Ion refers to a positively or negatively charged particle formed when an atom or group of atoms loses or gains electrons. |
| Proton1 | Subatomic particle with positive charge that is nucleus of hydrogen atom is called proton. |
| Temperature1 | Temperature refers to measure of hotness of object on a quantitative scale. In gases, proportional to average kinetic energy of molecules. |
| Equation1 | An equation is a mathematical expression with an equal sign in it. It signifies that the numerical or vector value on one side of the = is the same as the numerical or vector value on the other side. An equation may include variables and parameters. If any of the variables are rates of change, the equation is called a differential equation. |
| Ionization1 | The process by which a neutral atom or molecule acquires a positive or negative charge is called ionization. |
| Molecule1 | The smallest part of an element or compound that can exist on its own is called the molecule. |
| Liter1 | A metric system unit of volume, usually used for liquids is referred to as liter. |
| Weight1 | Force of gravity of an object is called weight. |
| Atom1 | Atom refers to the smallest particle of an element which can exist. |

31

## Chapter 3. Acid - Base Chemistry

| | |
|---|---|
| Normal1 | Perpendicular to plane of interest is called normal. |
| Charge1 | A property of atomic particles. Electrons and protons have opposite charges and attract each other. The charge on an electron is negative and that on a proton is positive. Measured in coulombs. |
| System1 | Defined collection of objects is called a system. |
| Absorption1 | The transfer of energy to a medium, such as body tissues, as a radiation beam passes through the medium is absorption. |
| Mole1 | The amount of substance containing $6 \times 10^{23}$ particles is the mole. |
| Matter1 | We call the commonly observed particles such as protons, neutrons and electrons matter particles, and their antiparticles, antimatter. |
| Image1 | Reproduction of object formed with lenses or mirrors is the image. |
| 2+b*x+c . | A function involving the second and lower power, and none higher, of the independent variable. A quadratic function may contain $x^2$ explicitly or it may contain terms like x*, where the second power of x is implied. In general a quadratic may be written as y=a*x |
| Activity1 | Number of decays per second of a radioactive substance is an activity. |
| Function1 | A mathematical function is a rule relating two sets of objects. Here we will restrict ourselves to objects that are numbers or vectors. One of the sets is called the domain of the function, the other is called the range of the function. Functions are frequently expressed as equations as for example Y=X+2. This function is interpreted as follows. For every X in the domain, add 2 to it to get the corresponding Y in the range. Because we are free to choose any X we want, X is called the independent variable. Because once X is chosen Y is fixed, we call Y the dependent variable. |
| Solids1 | A phase of matter with molecules that remain close to fixed equilibrium positions due to strong interactions between the molecules, resulting in the characteristic definite shape and definite volume are solids. |
| Negative1 | The sign of the electric charge on the electron is negative. |
| Gas1 | State of matter that expands to fill container is referred to as gas. |

## Chapter 3. Acid - Base Chemistry

| | |
|---|---|
| Mixture1 | Matter made of unlike parts that have a variable composition and can be separated into their component parts by physical means is a mixture. |
| Pressure1 | Force per unit area is referred to as the pressure. |
| Secondary1 | The output coil of a transformer is referred to as secondary. |
| Fluid1 | Fluid refers to material that flows, i.e. liquids, gases, and plasmas. |
| Dynamics1 | Study of motion of particles acted on by forces is called dynamics. |
| Electrolyte1 | Electrolyte refers to water solution of ionic substances that conducts an electric current . |

## Chapter 4. Organic and Biochemistry

| | |
|---|---|
| Atom1 | Atom refers to the smallest particle of an element which can exist. |
| Properties1 | Properties refers to qualities or attributes that, taken together, are usually unique to an object; for example, color, texture, and size. |
| Stable1 | Does not decay is stable. |
| Periodic table1 | Periodic table refers to a chart of all the known chemical elements arranged according to the number of protons in the nucleus . Elements with similar properties are grouped together in the same column. |
| Electronegativity1 | The comparative ability of atoms of an element to attract bonding electrons is referred to as electronegativity. |
| Molecule1 | The smallest part of an element or compound that can exist on its own is called the molecule. |
| Element1 | A pure substance that cannot be split up into anything simpler is called an element. |
| Energy1 | Energy refers to non-material property capable of causing changes in matter. |
| Melting1 | A solid changes to a liquid at its melting point. |
| Weight1 | Force of gravity of an object is called weight. |
| State1 | Dynamical systems evolve over the course of time. The state of the system at any instant may be identified by the values of certain variables at that instant. For example specifying the angle from the vertical and the velocity of a frictionless pendulum allows us to predict its position and velocity at any future time. Therefore the state of the pendulum at any instant is its position and velocity. In this example the position and velocity are known as state variables. |
| Cell1 | In electricity, a cell is a combination of metals and chemicals that produces a voltage and can cause a current. |
| Isomer1 | One of several nuclides with the same number of neutrons and protons capable of existing for a measurable time in different nuclear energy states is called an isomer. |
| Density1 | The mass of a given volume of substance. It has units of kg/m3 or g/cm3. When the density is high the particles are closely packed. |

## Chapter 4. Organic and Biochemistry

| | |
|---|---|
| Boiling point1 | Temperature at which a substance, under normal atmospheric pressure, changes from a liquid to a vapor state is the boiling point. |
| Pressure1 | Force per unit area is referred to as the pressure. |
| Liquid1 | Material that has fixed volume but whose shape depends on the container is called liquid. |
| Ion1 | Ion refers to a positively or negatively charged particle formed when an atom or group of atoms loses or gains electrons. |
| Unstable1 | Matter that is capable of undergoing spontaneous change, as in a radioactive nuclide or an excited nuclear system. An unstable particle is any elementary particle that spontaneously decays into other particles. |
| Resonance1 | When the frequency of an external force matches the natural frequency and standing waves are set up, we have resonance. |
| Secondary1 | The output coil of a transformer is referred to as secondary. |
| Cycle1 | In wave motion, one cycle is a trough and a crest for a transverse wave, or a compression and a rarefaction for a longitudinal wave. |
| Property1 | A characteristic that is inherently associated with the object which is said to have that property. For example the mass of an object is one of its properties. So also might be color, density and many other characteristics. Properties are classified as extensive or intensive. Extensive properties increase in proportion to the size of the object, as mass does for example. Intensive properties are independent of the size of the object. The density for examples remains the same if I cut an object in half and throw half of it away. Things like an object's position or velocity are not considered to be properties of the object. They are not a characteristic of the object only but are also dependent on the reference frame in which the object is located. |
| Function1 | A mathematical function is a rule relating two sets of objects. Here we will restrict ourselves to objects that are numbers or vectors. One of the sets is called the domain of the function, the other is called the range of the function. Functions are frequently expressed as equations as for example $Y=X+2$. This function is interpreted as follows. For every X in the domain, add 2 to it to get the corresponding Y in the range. Because we are free to choose any X we want, X is called the independent variable. Because once X is chosen Y is fixed, we call Y the dependent variable. |

## Chapter 4. Organic and Biochemistry

| | |
|---|---|
| Electron1 | Subatomic particle of small mass and negative charge found in every atom is called the electron. |
| Amp1 | The unit of electric current is referred to as amp. |
| Range1 | The range is the set of values that the dependent variable of a function may take on. A range may be finite as in the set of numbers {1, 2, 3.n} or infinite as in all the mumbers between 0 and 1. |
| Activity1 | Number of decays per second of a radioactive substance is an activity. |
| Conversion1 | Chemical process turning U308 into UF6 preparatory to enrichment is conversion. |
| Mole1 | The amount of substance containing $6 \times 10^{23}$ particles is the mole. |
| Normal1 | Perpendicular to plane of interest is called normal. |
| Equation1 | An equation is a mathematical expression with an equal sign in it. It signifies that the numerical or vector value on one side of the = is the same as the numerical or vector value on the other side. An equation may include variables and parameters. If any of the variables are rates of change, the equation is called a differential equation. |
| Velocity1 | The ratio of change in position with respect to the time interval over which the change occurred is referred to as velocity. |
| Units1 | The units one uses should be of a size that makes sense for the particular subject at hand. It is easiest to define units in each area of science and then relate them to one another than to go around measuring particle masses in grams or cheese in proton mass units. In particle physics the standard unit is the unit of energy gev. One ev is the amount of energy that an electron gains when it moves through a potential difference of 1 Volt . G stands for Giga, or $10^9$. Thus a gev is a billion electron Volts. The mass-energy of a proton or neutron is approximately 1 gev. |
| Slope1 | Ratio of the vertical separation, or rise to the horizontal separation, or run are called the slope. |
| Accuracy1 | Accuracy refers to closeness of a measurement to the standard value of that quantity. |
| Temperature1 | Temperature refers to measure of hotness of object on a quantitative scale. In gases, proportional to average kinetic energy of molecules. |

## Chapter 5. Analysis and Constituents in Water

| | |
|---|---|
| State1 | Dynamical systems evolve over the course of time. The state of the system at any instant may be identified by the values of certain variables at that instant. For example specifying the angle from the vertical and the velocity of a frictionless pendulum allows us to predict its position and velocity at any future time. Therefore the state of the pendulum at any instant is its position and velocity. In this example the position and velocity are known as state variables. |
| Current1 | Current refers to a flow of charge. Measured in amps. |
| Precision1 | Degree of exactness in a measurement is precision. |
| Accuracy1 | Accuracy refers to closeness of a measurement to the standard value of that quantity. |
| Weight1 | Force of gravity of an object is called weight. |
| Liter1 | A metric system unit of volume, usually used for liquids is referred to as liter. |
| Equilibrium1 | Equilibrium refers to condition in which net force is equal to zero. Condition in which net torque on object is zero. |
| Contamination1 | Radioactive material deposited or dispersed in materials or places where it is not wanted is referred to as contamination. |
| Color1 | The visual perception of light that enables human eyes to differentiate between wavelengths of the visible spectrum, with the longest wavelengths appearing red and the shortest appearing blue or violet is called color. |
| Equation1 | An equation is a mathematical expression with an equal sign in it. It signifies that the numerical or vector value on one side of the = is the same as the numerical or vector value on the other side. An equation may include variables and parameters. If any of the variables are rates of change, the equation is called a differential equation. |
| Range1 | The range is the set of values that the dependent variable of a function may take on. A range may be finite as in the set of numbers {1, 2, 3.n} or infinite as in all the mumbers between 0 and 1. |
| Mole1 | The amount of substance containing $6 \times 10^{23}$ particles is the mole. |
| Ion1 | Ion refers to a positively or negatively charged particle formed when an atom or group of atoms loses or gains electrons. |

## Chapter 5. Analysis and Constituents in Water

| | |
|---|---|
| Metal1 | Matter having the physical properties of conductivity, malleability, ductility, and luster is referred to as metal. |
| Charge1 | A property of atomic particles. Electrons and protons have opposite charges and attract each other. The charge on an electron is negative and that on a proton is positive. Measured in coulombs. |
| System1 | Defined collection of objects is called a system. |
| Cell1 | In electricity, a cell is a combination of metals and chemicals that produces a voltage and can cause a current. |
| Plastic1 | The behaviour of a material is plastic if it does not regain its original shape when the deforming force is removed. |
| Activity1 | Number of decays per second of a radioactive substance is an activity. |
| Electron1 | Subatomic particle of small mass and negative charge found in every atom is called the electron. |
| Meter1 | Meter refers to SI unit of length. |
| Evaporation1 | Change from liquid to vapor state are called the evaporation. |
| Potential difference1 | Difference in electric potential between two points is called the potential difference. |
| Period1 | Period refers to time needed to repeat one complete cycle of motion. |
| Voltage1 | Voltage refers to potential difference. It is a measure of the change in energy that one coulomb of electric charge undergoes when moved between 2 points. |
| Slope1 | Ratio of the vertical separation, or rise to the horizontal separation, or run are called the slope. |
| Light1 | Electromagnetic radiation with wavelengths between 400 and 700 nm that is visible is called light. |
| Intensity1 | The amount of radiation, for example, the number of photons arriving in a given time, is called intensity. |

## Chapter 5. Analysis and Constituents in Water

| | |
|---|---|
| Absorption1 | The transfer of energy to a medium, such as body tissues, as a radiation beam passes through the medium is absorption. |
| Distance1 | Distance refers to separation between two points. A scalar quantity. |
| Function1 | A mathematical function is a rule relating two sets of objects. Here we will restrict ourselves to objects that are numbers or vectors. One of the sets is called the domain of the function, the other is called the range of the function. Functions are frequently expressed as equations as for example $Y=X+2$. This function is interpreted as follows. For every X in the domain, add 2 to it to get the corresponding Y in the range. Because we are free to choose any X we want, X is called the independent variable. Because once X is chosen Y is fixed, we call Y the dependent variable. |
| Wavelength1 | Distance between corresponding points on two successive waves is referred to as wavelength. |
| Reflection1 | Reflection refers to the change when light, sound, or other waves bounce backwards off a boundary . |
| Attenuation1 | The process by which a compound is reduced in concentration over time, through adsorption, degradation, dilution, and/or transformation. Radiologically, it is the reduction of the intensity of radiation upon passage through a medium. The attenuation is caused by absorption and scattering. |
| Monochromatic light1 | Light of a single wavelength is called monochromatic light. |
| Detector1 | Detector refers to any device used to sense the passage of a particle; also a collection of such devices designed so that each serves a particular purpose in allowing physicists to reconstruct particle events. |
| Lens1 | Lens refers to optical device designed to converge or diverge light. |
| Diffraction1 | Bending of waves around object in their path is called diffraction. |
| Galvanometer1 | Device used to measure very small currents is a galvanometer. |
| Excited state1 | Energy level of atom higher than ground state is called the excited state. |

## Chapter 5. Analysis and Constituents in Water

| | |
|---|---|
| Matter1 | We call the commonly observed particles such as protons, neutrons and electrons matter particles, and their antiparticles, antimatter. |
| Interference1 | Phenomenon of light where the relative phase difference between two light waves produces light or dark spots, a result of light's wavelike nature is called the interference. |
| Solids1 | A phase of matter with molecules that remain close to fixed equilibrium positions due to strong interactions between the molecules, resulting in the characteristic definite shape and definite volume are solids. |
| Phase1 | The particles in a wave, which are in the same state of vibration, i.e., the same position and the same direction of motion are said to be in the same phase. |
| Solid1 | State of matter with fixed volume and shape is referred to as solid. |
| Molecule1 | The smallest part of an element or compound that can exist on its own is called the molecule. |
| Charged1 | Object that has an unbalance of positive and negative electrical charges is referred to as charged. |
| Particle1 | In 'particle physics', a subatomic object with definite mass and charge . |
| Set1 | In mathematics a set is a collection of related objects. The mathematical usage is similar to the ordinary English meaning of the word. The objects that make up a set are called the elements of the set. If a set contains an unlimited number of elements it is an infinite set. Otherwise it is a finite set. |
| Temperature1 | Temperature refers to measure of hotness of object on a quantitative scale. In gases, proportional to average kinetic energy of molecules. |
| Boiling point1 | Temperature at which a substance, under normal atmospheric pressure, changes from a liquid to a vapor state is the boiling point. |
| Movement1 | Change of position is called movement. |
| Liquid1 | Material that has fixed volume but whose shape depends on the container is called liquid. |
| Gas1 | State of matter that expands to fill container is referred to as gas. |

## Chapter 5. Analysis and Constituents in Water

| | |
|---|---|
| Infrared1 | A type of electromagnetic radiation with a wavelength longer than that of light is called infrared. |
| Atom1 | Atom refers to the smallest particle of an element which can exist. |
| Condensing1 | Condensing refers to changing a vapour into a liquid. This change is accompanied by a giving out of energy. |
| Time period1 | The time taken by a wave to travel through a distance equal to its wavelength is called its time period. |
| Waste1 | High-level waste is highly radioactive material arising from nuclear fission. It is recovered from reprocessing spent fuel, though some countries regard spent fuel itself as HLW and plan to dispose of it in that form. It requires very careful handling, storage and disposal. |
| Cycle1 | In wave motion, one cycle is a trough and a crest for a transverse wave, or a compression and a rarefaction for a longitudinal wave. |
| Pressure1 | Force per unit area is referred to as the pressure. |
| Gases1 | A phase of matter composed of molecules that are relatively far apart moving freely in a constant, random motion and have weak cohesive forces acting between them, resulting in the characteristic indefinite shape and indefinite volume of a gas are called the gases. |
| Significant figures1 | The number of digits in a numerical value that are reliably known. If the numbers being used in a calculation are measured values, there will always be a limit on the accuracy of the measurement. The results of any calculations based on those numbers should not be reported with more significant figures than the least acurate of the measured values. |
| Second1 | Second is a SI unit of time. |
| Conversion1 | Chemical process turning U308 into UF6 preparatory to enrichment is conversion. |
| Stable1 | Does not decay is stable. |
| Energy1 | Energy refers to non-material property capable of causing changes in matter. |
| Heat1 | Heat refers to quantity of energy transferred from one object to another because of a difference in temperature. |

## Chapter 5. Analysis and Constituents in Water

| | |
|---|---|
| Normal1 | Perpendicular to plane of interest is called normal. |
| Gravity1 | A force with infinite range which acts between objects, such as planets, according to their mass is called gravity. |
| Units1 | The units one uses should be of a size that makes sense for the particular subject at hand. It is easiest to define units in each area of science and then relate them to one another than to go around measuring particle masses in grams or cheese in proton mass units. In particle physics the standard unit is the unit of energy gev. One ev is the amount of energy that an electron gains when it moves through a potential difference of 1 Volt . G stands for Giga, or $10^9$. Thus a gev is a billion electron Volts. The mass-energy of a proton or neutron is approximately 1 gev. |
| Velocity1 | The ratio of change in position with respect to the time interval over which the change occurred is referred to as velocity. |
| Volt1 | The unit of voltage or potential difference is a volt. |

53

## Chapter 6. Microbiology

| | |
|---|---|
| Temperature1 | Temperature refers to measure of hotness of object on a quantitative scale. In gases, proportional to average kinetic energy of molecules. |
| Fission1 | The splitting of atomic nuclei into smaller particles is referred to as fission. |
| Cell1 | In electricity, a cell is a combination of metals and chemicals that produces a voltage and can cause a current. |
| Pressure1 | Force per unit area is referred to as the pressure. |
| Liquid1 | Material that has fixed volume but whose shape depends on the container is called liquid. |
| Weight1 | Force of gravity of an object is called weight. |
| Energy1 | Energy refers to non-material property capable of causing changes in matter. |
| State1 | Dynamical systems evolve over the course of time. The state of the system at any instant may be identified by the values of certain variables at that instant. For example specifying the angle from the vertical and the velocity of a frictionless pendulum allows us to predict its position and velocity at any future time. Therefore the state of the pendulum at any instant is its position and velocity. In this example the position and velocity are known as state variables. |
| Heat1 | Heat refers to quantity of energy transferred from one object to another because of a difference in temperature. |
| Range1 | The range is the set of values that the dependent variable of a function may take on. A range may be finite as in the set of numbers {1, 2, 3.n} or infinite as in all the mumbers between 0 and 1. |
| Matter1 | We call the commonly observed particles such as protons, neutrons and electrons matter particles, and their antiparticles, antimatter. |
| Sun1 | Sun refers to the star at the centre of the solar system. The star the Earth orbits. |
| Switch1 | Switch refers to a device in an electrical circuit that breaks or makes a complete path for the current. |
| Nucleus1 | Nucleus refers to the central part of an atom. It contains all the positive charge and almost all the mass of the atom. |

## Chapter 6. Microbiology

| | |
|---|---|
| Conversion1 | Chemical process turning U308 into UF6 preparatory to enrichment is conversion. |
| Magnitude1 | Magnitude refers to the size of a thing, without regard for its sign or direction. Similar to the absolute value of a number but applies to vectors as well. |
| Color1 | The visual perception of light that enables human eyes to differentiate between wavelengths of the visible spectrum, with the longest wavelengths appearing red and the shortest appearing blue or violet is called color. |
| Light1 | Electromagnetic radiation with wavelengths between 400 and 700 nm that is visible is called light. |
| Decay1 | Any process in which a particle disappears and in its place two or more different particles appear is decay. |
| Enrichment1 | Physical process of increasing the proportion of U-235 to U-238 is enrichment. |
| Quantity1 | A numerical value either scalar or vector, which describes some attribute of an object like its position or its velocity . We sometimes speak of physical quantities to signify that we are talking about an object's properties or attributes as opposed to a purely mathematical quantity. |
| Cycle1 | In wave motion, one cycle is a trough and a crest for a transverse wave, or a compression and a rarefaction for a longitudinal wave. |
| Molecule1 | The smallest part of an element or compound that can exist on its own is called the molecule. |
| Particle1 | In 'particle physics', a subatomic object with definite mass and charge . |
| Second1 | Second is a SI unit of time. |
| Daughter1 | A nucleus formed by the radioactive decay of a different nuclide is called the daughter. |
| Element1 | A pure substance that cannot be split up into anything simpler is called an element. |
| Retardation1 | Negative acceleration is called retardation. In retardation the velocity of a body decreases with time. |
| Phase1 | The particles in a wave, which are in the same state of vibration, i.e., the same position and the same direction of motion are said to be in the same phase. |

## Chapter 6. Microbiology

| | |
|---|---|
| Period1 | Period refers to time needed to repeat one complete cycle of motion. |
| Equation1 | An equation is a mathematical expression with an equal sign in it. It signifies that the numerical or vector value on one side of the = is the same as the numerical or vector value on the other side. An equation may include variables and parameters. If any of the variables are rates of change, the equation is called a differential equation. |
| Slope1 | Ratio of the vertical separation, or rise to the horizontal separation, or run are called the slope. |
| Units1 | The units one uses should be of a size that makes sense for the particular subject at hand. It is easiest to define units in each area of science and then relate them to one another than to go around measuring particle masses in grams or cheese in proton mass units. In particle physics the standard unit is the unit of energy gev. One ev is the amount of energy that an electron gains when it moves through a potential difference of 1 Volt . G stands for Giga, or $10^9$. Thus a gev is a billion electron Volts. The mass-energy of a proton or neutron is approximately 1 gev. |
| Waste1 | High-level waste is highly radioactive material arising from nuclear fission. It is recovered from reprocessing spent fuel, though some countries regard spent fuel itself as HLW and plan to dispose of it in that form. It requires very careful handling, storage and disposal. |
| Solid1 | State of matter with fixed volume and shape is referred to as solid. |
| Solids1 | A phase of matter with molecules that remain close to fixed equilibrium positions due to strong interactions between the molecules, resulting in the characteristic definite shape and definite volume are solids. |
| Radiation1 | Electromagnetic waves that carry energy are referred to as radiation. |
| Gas1 | State of matter that expands to fill container is referred to as gas. |
| Ultraviolet1 | Ultraviolet refers to electromagnetic waves with a wavelength shorter than that of visible light. |
| Contamination1 | Radioactive material deposited or dispersed in materials or places where it is not wanted is referred to as contamination. |
| Plates1 | Large sections of the Earth's crust which float on the mantle are called plates. |
| Secondary1 | The output coil of a transformer is referred to as secondary. |

## Chapter 6. Microbiology

| | |
|---|---|
| Negative1 | The sign of the electric charge on the electron is negative. |
| Intensity1 | The amount of radiation, for example, the number of photons arriving in a given time, is called intensity. |
| Activity1 | Number of decays per second of a radioactive substance is an activity. |
| Density1 | The mass of a given volume of substance. It has units of kg/m3 or g/cm3. When the density is high the particles are closely packed. |
| Set1 | In mathematics a set is a collection of related objects. The mathematical usage is similar to the ordinary English meaning of the word. The objects that make up a set are called the elements of the set. If a set contains an unlimited number of elements it is an infinite set. Otherwise it is a finite set. |
| Event1 | An event occurs when two particles collide or a single particle decay . Particle theories predict the probabilities of various events occurring when many similar collisions or decays are studied. They cannot predict the outcome for a single collision or decay. |
| Frequency1 | Frequency refers to number of occurrences per unit time. |
| Accuracy1 | Accuracy refers to closeness of a measurement to the standard value of that quantity. |
| Vacuum1 | Vacuum refers to a region of space containing no matter. In practice, a region of gas at very low pressure. |
| Core1 | In electromagnetism, the material inside the coils of a transformer or electromagnet is a core. |
| Eluant1 | A washing solution is called eluant. |
| Efficiency1 | Ratio of output work to input work is called efficiency. |
| Electron1 | Subatomic particle of small mass and negative charge found in every atom is called the electron. |
| Force1 | Force refers to agent that results in accelerating or deforming an object. |

## Chapter 6. Microbiology

Function1

A mathematical function is a rule relating two sets of objects. Here we will restrict ourselves to objects that are numbers or vectors. One of the sets is called the domain of the function, the other is called the range of the function. Functions are frequently expressed as equations as for example Y=X+2. This function is interpreted as follows. For every X in the domain, add 2 to it to get the corresponding Y in the range. Because we are free to choose any X we want, X is called the independent variable. Because once X is chosen Y is fixed, we call Y the dependent variable.

Liter1

A metric system unit of volume, usually used for liquids is referred to as liter.

## Chapter 7. Water, Wastes, and Disease

| | |
|---|---|
| Waste1 | High-level waste is highly radioactive material arising from nuclear fission. It is recovered from reprocessing spent fuel, though some countries regard spent fuel itself as HLW and plan to dispose of it in that form. It requires very careful handling, storage and disposal. |
| Period1 | Period refers to time needed to repeat one complete cycle of motion. |
| State1 | Dynamical systems evolve over the course of time. The state of the system at any instant may be identified by the values of certain variables at that instant. For example specifying the angle from the vertical and the velocity of a frictionless pendulum allows us to predict its position and velocity at any future time. Therefore the state of the pendulum at any instant is its position and velocity. In this example the position and velocity are known as state variables. |
| System1 | Defined collection of objects is called a system. |
| Fluids1 | Fluids refers to matter that has the ability to flow or be poured; the individual molecules of a fluid are able to move, rolling over or by one another . |
| Fluid1 | Fluid refers to material that flows, i.e. liquids, gases, and plasmas. |
| Units1 | The units one uses should be of a size that makes sense for the particular subject at hand. It is easiest to define units in each area of science and then relate them to one another than to go around measuring particle masses in grams or cheese in proton mass units. In particle physics the standard unit is the unit of energy gev. One ev is the amount of energy that an electron gains when it moves through a potential difference of 1 Volt . G stands for Giga, or $10^9$. Thus a gev is a billion electron Volts. The mass-energy of a proton or neutron is approximately 1 gev. |
| Cell1 | In electricity, a cell is a combination of metals and chemicals that produces a voltage and can cause a current. |
| Function1 | A mathematical function is a rule relating two sets of objects. Here we will restrict ourselves to objects that are numbers or vectors. One of the sets is called the domain of the function, the other is called the range of the function. Functions are frequently expressed as equations as for example $Y=X+2$. This function is interpreted as follows. For every X in the domain, add 2 to it to get the corresponding Y in the range. Because we are free to choose any X we want, X is called the independent variable. Because once X is chosen Y is fixed, we call Y the dependent variable. |
| Stable1 | Does not decay is stable. |

## Chapter 7. Water, Wastes, and Disease

| | |
|---|---|
| Cycle1 | In wave motion, one cycle is a trough and a crest for a transverse wave, or a compression and a rarefaction for a longitudinal wave. |
| Second1 | Second is a SI unit of time. |
| Range1 | The range is the set of values that the dependent variable of a function may take on. A range may be finite as in the set of numbers {1, 2, 3.n} or infinite as in all the mumbers between 0 and 1. |
| Contamination1 | Radioactive material deposited or dispersed in materials or places where it is not wanted is referred to as contamination. |
| Matter1 | We call the commonly observed particles such as protons, neutrons and electrons matter particles, and their antiparticles, antimatter. |
| Origin1 | Origin refers to the point in a reference frame from which measurements are made. It is the location of the zero value for each axis in the frame. |
| Current1 | Current refers to a flow of charge. Measured in amps. |
| Weight1 | Force of gravity of an object is called weight. |
| Density1 | The mass of a given volume of substance. It has units of kg/m3 or g/cm3. When the density is high the particles are closely packed. |
| Cow1 | A radioisotope generator system is referred to as cow. |
| Normal1 | Perpendicular to plane of interest is called normal. |
| Gas1 | State of matter that expands to fill container is referred to as gas. |
| Frequency1 | Frequency refers to number of occurrences per unit time. |
| Temperature1 | Temperature refers to measure of hotness of object on a quantitative scale. In gases, proportional to average kinetic energy of molecules. |
| Negative1 | The sign of the electric charge on the electron is negative. |
| Movement1 | Change of position is called movement. |

## Chapter 7. Water, Wastes, and Disease

| Color1 | The visual perception of light that enables human eyes to differentiate between wavelengths of the visible spectrum, with the longest wavelengths appearing red and the shortest appearing blue or violet is called color. |
|---|---|
| Precision1 | Degree of exactness in a measurement is precision. |
| Stress1 | Stress is defined as force per unit area. This is one of the most basic engineering quantities. |
| Strain1 | Strain is the change in length per unit length. It is normally computed as (Lf - Lo) / Lo where Lf is the final length and Lo is the initial length. When testing materials, a gage length is normally specified known; this represents Lo. |
| Barrier1 | Radiation-absorbing material, such as lead or concrete, used to reduce radiation exposure. A primary barrier attenuates useful beam to the required degree. A secondary barrier attenuates stray radiation to the required degree. |
| Decay1 | Any process in which a particle disappears and in its place two or more different particles appear is decay. |
| Radiation1 | Electromagnetic waves that carry energy are referred to as radiation. |
| Ultraviolet1 | Ultraviolet refers to electromagnetic waves with a wavelength shorter than that of visible light. |
| Attenuation1 | The process by which a compound is reduced in concentration over time, through adsorption, degradation, dilution, and/or transformation. Radiologically, it is the reduction of the intensity of radiation upon passage through a medium. The attenuation is caused by absorption and scattering. |
| Intensity1 | The amount of radiation, for example, the number of photons arriving in a given time, is called intensity. |
| Equation1 | An equation is a mathematical expression with an equal sign in it. It signifies that the numerical or vector value on one side of the = is the same as the numerical or vector value on the other side. An equation may include variables and parameters. If any of the variables are rates of change, the equation is called a differential equation. |
| Groundwater1 | Groundwater refers to water found in the voids or free space of soils and rocks underground. |
| Secondary1 | The output coil of a transformer is referred to as secondary. |

## Chapter 7. Water, Wastes, and Disease

| | |
|---|---|
| Resistance1 | Ratio of potential difference across device to current through it are called the resistance. |
| Event1 | An event occurs when two particles collide or a single particle decay . Particle theories predict the probabilities of various events occurring when many similar collisions or decays are studied. They cannot predict the outcome for a single collision or decay. |
| Solids1 | A phase of matter with molecules that remain close to fixed equilibrium positions due to strong interactions between the molecules, resulting in the characteristic definite shape and definite volume are solids. |
| Liter1 | A metric system unit of volume, usually used for liquids is referred to as liter. |
| Sun1 | Sun refers to the star at the centre of the solar system. The star the Earth orbits. |
| Solid1 | State of matter with fixed volume and shape is referred to as solid. |

## Chapter 8. Water Components and Quality Standards

| | |
|---|---|
| Mercury1 | Mercury refers to the innermost planet in the solar system, and a metallic element that is liquid at room temperature. |
| Element1 | A pure substance that cannot be split up into anything simpler is called an element. |
| Ionized1 | An atom or a particle that has a net charge because it has gained or lost electrons is ionized. |
| System1 | Defined collection of objects is called a system. |
| Contamination1 | Radioactive material deposited or dispersed in materials or places where it is not wanted is referred to as contamination. |
| Period1 | Period refers to time needed to repeat one complete cycle of motion. |
| Metal1 | Matter having the physical properties of conductivity, malleability, ductility, and luster is referred to as metal. |
| Normal1 | Perpendicular to plane of interest is called normal. |
| Solid1 | State of matter with fixed volume and shape is referred to as solid. |
| Waste1 | High-level waste is highly radioactive material arising from nuclear fission. It is recovered from reprocessing spent fuel, though some countries regard spent fuel itself as HLW and plan to dispose of it in that form. It requires very careful handling, storage and disposal. |
| Groundwater1 | Groundwater refers to water found in the voids or free space of soils and rocks underground. |
| Range1 | The range is the set of values that the dependent variable of a function may take on. A range may be finite as in the set of numbers {1, 2, 3.n} or infinite as in all the mumbers between 0 and 1. |
| Ion1 | Ion refers to a positively or negatively charged particle formed when an atom or group of atoms loses or gains electrons. |
| Absorption1 | The transfer of energy to a medium, such as body tissues, as a radiation beam passes through the medium is absorption. |
| Stable1 | Does not decay is stable. |

## Chapter 8. Water Components and Quality Standards

| | |
|---|---|
| Isotope1 | Atomic nuclei having same number of protons but different numbers of neutrons are called an isotope. |
| Radioactive1 | Radioactive nuclei are unstable. They decay by emitting alpha or beta particles or gamma rays. |
| Uranium1 | A mildly radioactive element with two isotopes which are fissile and two, which are fertile . Uranium is the basic raw material of nuclear energy. |
| Weight1 | Force of gravity of an object is called weight. |
| Matter1 | We call the commonly observed particles such as protons, neutrons and electrons matter particles, and their antiparticles, antimatter. |
| Molecule1 | The smallest part of an element or compound that can exist on its own is called the molecule. |
| Cell1 | In electricity, a cell is a combination of metals and chemicals that produces a voltage and can cause a current. |
| Strain1 | Strain is the change in length per unit length. It is normally computed as $(Lf - Lo) / Lo$ where $Lf$ is the final length and $Lo$ is the initial length. When testing materials, a gage length is normally specified known; this represents $Lo$. |
| State1 | Dynamical systems evolve over the course of time. The state of the system at any instant may be identified by the values of certain variables at that instant. For example specifying the angle from the vertical and the velocity of a frictionless pendulum allows us to predict its position and velocity at any future time. Therefore the state of the pendulum at any instant is its position and velocity. In this example the position and velocity are known as state variables. |
| Current1 | Current refers to a flow of charge. Measured in amps. |
| Light1 | Electromagnetic radiation with wavelengths between 400 and 700 nm that is visible is called light. |
| Activity1 | Number of decays per second of a radioactive substance is an activity. |
| Radiation1 | Electromagnetic waves that carry energy are referred to as radiation. |

## Chapter 8. Water Components and Quality Standards

| | |
|---|---|
| Units1 | The units one uses should be of a size that makes sense for the particular subject at hand. It is easiest to define units in each area of science and then relate them to one another than to go around measuring particle masses in grams or cheese in proton mass units. In particle physics the standard unit is the unit of energy gev. One ev is the amount of energy that an electron gains when it moves through a potential difference of 1 Volt . G stands for Giga, or $10^9$. Thus a gev is a billion electron Volts. The mass-energy of a proton or neutron is approximately 1 gev. |
| Second1 | Second is a SI unit of time. |
| Intensity1 | The amount of radiation, for example, the number of photons arriving in a given time, is called intensity. |
| Quantity1 | A numerical value either scalar or vector, which describes some attribute of an object like its position or its velocity . We sometimes speak of physical quantities to signify that we are talking about an object's properties or attributes as opposed to a purely mathematical quantity. |
| Electricity1 | Electricity refers to a general term that describes the presence of a voltage or a current. |
| Pressure1 | Force per unit area is referred to as the pressure. |
| Energy1 | Energy refers to non-material property capable of causing changes in matter. |
| Sievert1 | Sievert is the derived SI unit of dose equivalent, defined as the absorbed dose of ionizing radiation multiplied by internationally-agreed-upon dimensionless weights, since different types of ionizing radiation cause different types of damage in living tissue. |
| Dose1 | More specifically referred to as 'absorbed dose', this is a measure of the energy deposited within a given mass of a patient. Absorbed dose is quantified by the unit called the 'rad'. |
| Rad1 | One rad is equal to an energy absorption of 100 ergs in a gram of any material. An 'erg' is a unit for quantifying energy . |
| Radioactivity1 | The property of spontaneously emitting alpha, beta, and/or gamma radiation as a result of nuclear disintegration is radioactivity. |
| Rads1 | Rads refers to a unit to measure the absorption of radiation by the body. A rad is equivalent to 100 ergs of energy from ionising radiation absorbed per gram of soft tissue. |

## Chapter 8. Water Components and Quality Standards

| | |
|---|---|
| Isotopes1 | Atoms with the same atomic number, but different mass numbers are called isotopes. |
| Power1 | Power refers to rate of doing work; rate of energy conversion. |
| Distance1 | Distance refers to separation between two points. A scalar quantity. |
| Particle1 | In 'particle physics', a subatomic object with definite mass and charge . |
| Beta1 | A type of nuclear radiation consisting of fast-moving electrons emitted from nuclei when they undergo radioactive decay is beta. |
| Gamma rays1 | Gamma rays are electromagnetic waves or photons emitted from the nucleus of an atom. |
| Radon1 | A heavy radioactive gas given off by rocks containing radium is called radon. |
| Dose equivalent1 | Dose equivalent refers to a parameter used to express the risk of the deleterious effects of ionization radiation upon living organisms. For radiation protection purposes, the quantity of the effective irradiation incurred by exposed persons, measured on a common scale in sievert or rem . |
| Equation1 | An equation is a mathematical expression with an equal sign in it. It signifies that the numerical or vector value on one side of the = is the same as the numerical or vector value on the other side. An equation may include variables and parameters. If any of the variables are rates of change, the equation is called a differential equation. |
| Vapor1 | The gaseous state of a substance that is normally in the liquid state is vapor. |
| Set1 | In mathematics a set is a collection of related objects. The mathematical usage is similar to the ordinary English meaning of the word. The objects that make up a set are called the elements of the set. If a set contains an unlimited number of elements it is an infinite set. Otherwise it is a finite set. |
| Frequency1 | Frequency refers to number of occurrences per unit time. |
| Origin1 | Origin refers to the point in a reference frame from which measurements are made. It is the location of the zero value for each axis in the frame. |
| Properties1 | Properties refers to qualities or attributes that, taken together, are usually unique to an object; for example, color, texture, and size. |

## Chapter 8. Water Components and Quality Standards

| | |
|---|---|
| Color1 | The visual perception of light that enables human eyes to differentiate between wavelengths of the visible spectrum, with the longest wavelengths appearing red and the shortest appearing blue or violet is called color. |
| Solids1 | A phase of matter with molecules that remain close to fixed equilibrium positions due to strong interactions between the molecules, resulting in the characteristic definite shape and definite volume are solids. |
| Temperature1 | Temperature refers to measure of hotness of object on a quantitative scale. In gases, proportional to average kinetic energy of molecules. |
| Secondary1 | The output coil of a transformer is referred to as secondary. |
| Vacuum1 | Vacuum refers to a region of space containing no matter. In practice, a region of gas at very low pressure. |
| Gravity1 | A force with infinite range which acts between objects, such as planets, according to their mass is called gravity. |
| Charged1 | Object that has an unbalance of positive and negative electrical charges is referred to as charged. |
| Sun1 | Sun refers to the star at the centre of the solar system. The star the Earth orbits. |
| Slope1 | Ratio of the vertical separation, or rise to the horizontal separation, or run are called the slope. |
| Event1 | An event occurs when two particles collide or a single particle decay . Particle theories predict the probabilities of various events occurring when many similar collisions or decays are studied. They cannot predict the outcome for a single collision or decay. |
| Time period1 | The time taken by a wave to travel through a distance equal to its wavelength is called its time period. |

## Chapter 8. Water Components and Quality Standards

| | |
|---|---|
| Function1 | A mathematical function is a rule relating two sets of objects. Here we will restrict ourselves to objects that are numbers or vectors. One of the sets is called the domain of the function, the other is called the range of the function. Functions are frequently expressed as equations as for example Y=X+2. This function is interpreted as follows. For every X in the domain, add 2 to it to get the corresponding Y in the range. Because we are free to choose any X we want, X is called the independent variable. Because once X is chosen Y is fixed, we call Y the dependent variable. |
| Thermometer1 | Device used to measure temperature is a thermometer. |
| Track1 | The record of the path of a particle traversing a detector is a track. |
| Fault1 | Breaks in the ground where plates join, e.g. San Andreas Fault in California. |

## Chapter 9. Water and Wastewater Treatment Operations

| | |
|---|---|
| Efficiency1 | Ratio of output work to input work is called efficiency. |
| Units1 | The units one uses should be of a size that makes sense for the particular subject at hand. It is easiest to define units in each area of science and then relate them to one another than to go around measuring particle masses in grams or cheese in proton mass units. In particle physics the standard unit is the unit of energy gev. One ev is the amount of energy that an electron gains when it moves through a potential difference of 1 Volt . G stands for Giga, or $10^9$. Thus a gev is a billion electron Volts. The mass-energy of a proton or neutron is approximately 1 gev. |
| Quantity1 | A numerical value either scalar or vector, which describes some attribute of an object like its position or its velocity . We sometimes speak of physical quantities to signify that we are talking about an object's properties or attributes as opposed to a purely mathematical quantity. |
| Heat1 | Heat refers to quantity of energy transferred from one object to another because of a difference in temperature. |
| Gases1 | A phase of matter composed of molecules that are relatively far apart moving freely in a constant, random motion and have weak cohesive forces acting between them, resulting in the characteristic indefinite shape and indefinite volume of a gas are called the gases. |
| Matter1 | We call the commonly observed particles such as protons, neutrons and electrons matter particles, and their antiparticles, antimatter. |
| Pressure1 | Force per unit area is referred to as the pressure. |
| Conversion1 | Chemical process turning U308 into UF6 preparatory to enrichment is conversion. |
| Unstable1 | Matter that is capable of undergoing spontaneous change, as in a radioactive nuclide or an excited nuclear system. An unstable particle is any elementary particle that spontaneously decays into other particles. |
| Gas1 | State of matter that expands to fill container is referred to as gas. |
| Ion1 | Ion refers to a positively or negatively charged particle formed when an atom or group of atoms loses or gains electrons. |
| Stable1 | Does not decay is stable. |

## Chapter 9. Water and Wastewater Treatment Operations

| | |
|---|---|
| Ultraviolet1 | Ultraviolet refers to electromagnetic waves with a wavelength shorter than that of visible light. |
| Light1 | Electromagnetic radiation with wavelengths between 400 and 700 nm that is visible is called light. |
| Radiation1 | Electromagnetic waves that carry energy are referred to as radiation. |
| Dose1 | More specifically referred to as 'absorbed dose', this is a measure of the energy deposited within a given mass of a patient. Absorbed dose is quantified by the unit called the 'rad'. |
| System1 | Defined collection of objects is called a system. |
| Solids1 | A phase of matter with molecules that remain close to fixed equilibrium positions due to strong interactions between the molecules, resulting in the characteristic definite shape and definite volume are solids. |
| Periodic1 | A dynamical system or function which at some point returns to the same state or value. If a system or function ever revisits the identical state or value it will continue to come back to it again and again in equal intervals of time. That is why we call such a system periodic. |
| Gravity1 | A force with infinite range which acts between objects, such as planets,according to their mass is called gravity. |
| Particle1 | In 'particle physics', a subatomic object with definite mass and charge . |
| Hydraulic1 | Hydraulic refers to using a fluid as a method of transmitting pressure. It allows forces to be magnified. |
| Weight1 | Force of gravity of an object is called weight. |
| Charged1 | Object that has an unbalance of positive and negative electrical charges is referred to as charged. |
| Crystalline1 | A regularly repeated crystal-like substructure is crystalline. |
| Velocity1 | The ratio of change in position with respect to the time interval over which the change occurred is referred to as velocity. |

## Chapter 9. Water and Wastewater Treatment Operations

| | |
|---|---|
| State1 | Dynamical systems evolve over the course of time. The state of the system at any instant may be identified by the values of certain variables at that instant. For example specifying the angle from the vertical and the velocity of a frictionless pendulum allows us to predict its position and velocity at any future time. Therefore the state of the pendulum at any instant is its position and velocity. In this example the position and velocity are known as state variables. |
| Electric current1 | The flow of electric charge electric field force field produced by an electrical charge is referred to as electric current. |
| Current1 | Current refers to a flow of charge. Measured in amps. |
| Force1 | Force refers to agent that results in accelerating or deforming an object. |
| Liquid1 | Material that has fixed volume but whose shape depends on the container is called liquid. |
| Barrier1 | Radiation-absorbing material, such as lead or concrete, used to reduce radiation exposure. A primary barrier attenuates useful beam to the required degree. A secondary barrier attenuates stray radiation to the required degree. |
| Negative1 | The sign of the electric charge on the electron is negative. |
| Cell1 | In electricity, a cell is a combination of metals and chemicals that produces a voltage and can cause a current. |
| Mercury1 | Mercury refers to the innermost planet in the solar system, and a metallic element that is liquid at room temperature. |
| Secondary1 | The output coil of a transformer is referred to as secondary. |
| Color1 | The visual perception of light that enables human eyes to differentiate between wavelengths of the visible spectrum, with the longest wavelengths appearing red and the shortest appearing blue or violet is called color. |
| Radon1 | A heavy radioactive gas given off by rocks containing radium is called radon. |
| Uranium1 | A mildly radioactive element with two isotopes which are fissile and two, which are fertile . Uranium is the basic raw material of nuclear energy. |

## Chapter 9. Water and Wastewater Treatment Operations

| | |
|---|---|
| Waste1 | High-level waste is highly radioactive material arising from nuclear fission. It is recovered from reprocessing spent fuel, though some countries regard spent fuel itself as HLW and plan to dispose of it in that form. It requires very careful handling, storage and disposal. |
| Groundwater1 | Groundwater refers to water found in the voids or free space of soils and rocks underground. |
| Solid1 | State of matter with fixed volume and shape is referred to as solid. |
| Supersaturated1 | Containing more than the normal saturation amount of a solute at a given temperature is supersaturated. |
| Plates1 | Large sections of the Earth's crust which float on the mantle are called plates. |
| Inertia1 | Inertia refers to tendency of object not to change its motion. |
| Frequency1 | Frequency refers to number of occurrences per unit time. |
| Contamination1 | Radioactive material deposited or dispersed in materials or places where it is not wanted is referred to as contamination. |
| Vacuum1 | Vacuum refers to a region of space containing no matter. In practice, a region of gas at very low pressure. |
| Range1 | The range is the set of values that the dependent variable of a function may take on. A range may be finite as in the set of numbers {1, 2, 3.n} or infinite as in all the mumbers between 0 and 1. |
| Energy1 | Energy refers to non-material property capable of causing changes in matter. |
| Function1 | A mathematical function is a rule relating two sets of objects. Here we will restrict ourselves to objects that are numbers or vectors. One of the sets is called the domain of the function, the other is called the range of the function. Functions are frequently expressed as equations as for example Y=X+2. This function is interpreted as follows. For every X in the domain, add 2 to it to get the corresponding Y in the range. Because we are free to choose any X we want, X is called the independent variable. Because once X is chosen Y is fixed, we call Y the dependent variable. |
| Period1 | Period refers to time needed to repeat one complete cycle of motion. |

## Chapter 9. Water and Wastewater Treatment Operations

| | |
|---|---|
| Equation1 | An equation is a mathematical expression with an equal sign in it. It signifies that the numerical or vector value on one side of the = is the same as the numerical or vector value on the other side. An equation may include variables and parameters. If any of the variables are rates of change, the equation is called a differential equation. |
| Acceleration1 | Change in velocity divided by time interval over which it occurred is an acceleration. |
| Density1 | The mass of a given volume of substance. It has units of kg/m3 or g/cm3. When the density is high the particles are closely packed. |
| Fluid1 | Fluid refers to material that flows, i.e. liquids, gases, and plasmas. |
| Friction1 | Force opposing relative motion of two objects are in contact is friction. |
| Viscosity1 | A measure of how easy or otherwise a liquid can be poured. A high viscosity liquid flows very slowly. |
| Slope1 | Ratio of the vertical separation, or rise to the horizontal separation, or run are called the slope. |
| Free fall1 | The motion of a body towards the earth when no other force except the force of gravity acts on it is called free fall. All freely falling bodies are weightless. |
| Mechanics1 | The study of objects in motion. Mechanics is normally limited to a small number of large slow objects, as opposed to statistical mechanics which deals with large numbers of objects, relativistic mechanics which deals with objects moving near the speed of light and quantum mechanics which deals with objects more or less the size of atoms. Mechanics encompasses the topics of kinematics and dynamics . |
| Crest1 | The point of maximum positive displacement on a transverse wave is called a crest. |
| Accuracy1 | Accuracy refers to closeness of a measurement to the standard value of that quantity. |
| Bernoulli's equation1 | In an irrotational fluid, the sum of the static pressure, the weight of the fluid per unit mass times the height, and half the density times the velocity squared is constant throughout the fluid is called Bernoulli's equation. |
| Jet1 | The name physicists give to a cluster of particles emerging from a collision or decay event all traveling in roughly the same direction and carrying a significant fraction of the energy in the event. The particles in the jet are chiefly hadrons . |

| Temperature1 | Temperature refers to measure of hotness of object on a quantitative scale. In gases, proportional to average kinetic energy of molecules. |

## Chapter 10. Mass Balances and Hydraulic Flow Regimes

| | |
|---|---|
| Hydraulic1 | Hydraulic refers to using a fluid as a method of transmitting pressure. It allows forces to be magnified. |
| Equation1 | An equation is a mathematical expression with an equal sign in it. It signifies that the numerical or vector value on one side of the = is the same as the numerical or vector value on the other side. An equation may include variables and parameters. If any of the variables are rates of change, the equation is called a differential equation. |
| State1 | Dynamical systems evolve over the course of time. The state of the system at any instant may be identified by the values of certain variables at that instant. For example specifying the angle from the vertical and the velocity of a frictionless pendulum allows us to predict its position and velocity at any future time. Therefore the state of the pendulum at any instant is its position and velocity. In this example the position and velocity are known as state variables. |
| Convection1 | Heat transfer by means of motion of fluid is a convection. |
| Dispersion1 | The splitting of light into its constituent colours is called dispersion. |
| Displacement1 | Displacement refers to change in position. A vector quantity. |
| Liquid1 | Material that has fixed volume but whose shape depends on the container is called liquid. |
| Fluid1 | Fluid refers to material that flows, i.e. liquids, gases, and plasmas. |
| Function1 | A mathematical function is a rule relating two sets of objects. Here we will restrict ourselves to objects that are numbers or vectors. One of the sets is called the domain of the function, the other is called the range of the function. Functions are frequently expressed as equations as for example $Y=X+2$. This function is interpreted as follows. For every X in the domain, add 2 to it to get the corresponding Y in the range. Because we are free to choose any X we want, X is called the independent variable. Because once X is chosen Y is fixed, we call Y the dependent variable. |
| Turbulence1 | Unstable and disorderly motion, as when a smooth, flowing stream becomes a churning rapid is referred to as turbulence. |
| Random1 | With no set order or pattern, we have random. |
| Movement1 | Change of position is called movement. |

97

## Chapter 10. Mass Balances and Hydraulic Flow Regimes

| | |
|---|---|
| Matter1 | We call the commonly observed particles such as protons, neutrons and electrons matter particles, and their antiparticles, antimatter. |
| Diffusion1 | The spreading out of a substance, due to the kinetic energy of its particles, to fill all of the available space are called diffusion. |
| Gas1 | State of matter that expands to fill container is referred to as gas. |
| Temperature1 | Temperature refers to measure of hotness of object on a quantitative scale. In gases, proportional to average kinetic energy of molecules. |
| System1 | Defined collection of objects is called a system. |
| Distance1 | Distance refers to separation between two points. A scalar quantity. |
| Period1 | Period refers to time needed to repeat one complete cycle of motion. |
| Units1 | The units one uses should be of a size that makes sense for the particular subject at hand. It is easiest to define units in each area of science and then relate them to one another than to go around measuring particle masses in grams or cheese in proton mass units. In particle physics the standard unit is the unit of energy gev. One ev is the amount of energy that an electron gains when it moves through a potential difference of 1 Volt . G stands for Giga, or $10^9$. Thus a gev is a billion electron Volts. The mass-energy of a proton or neutron is approximately 1 gev. |
| Negative1 | The sign of the electric charge on the electron is negative. |
| Efficiency1 | Ratio of output work to input work is called efficiency. |
| Particle1 | In 'particle physics', a subatomic object with definite mass and charge . |
| Longitudinal1 | A type of wave motion in which the oscillations are parallel to the direction of wave travel is called longitudinal. |
| Barrier1 | Radiation-absorbing material, such as lead or concrete, used to reduce radiation exposure. A primary barrier attenuates useful beam to the required degree. A secondary barrier attenuates stray radiation to the required degree. |
| Element1 | A pure substance that cannot be split up into anything simpler is called an element. |

## Chapter 10. Mass Balances and Hydraulic Flow Regimes

| | |
|---|---|
| Velocity1 | The ratio of change in position with respect to the time interval over which the change occurred is referred to as velocity. |
| Time period1 | The time taken by a wave to travel through a distance equal to its wavelength is called its time period. |
| Decay1 | Any process in which a particle disappears and in its place two or more different particles appear is decay. |
| Second1 | Second is a SI unit of time. |
| Gradient1 | Gradient refers to the slope of a graph. |
| Cycle1 | In wave motion, one cycle is a trough and a crest for a transverse wave, or a compression and a rarefaction for a longitudinal wave. |
| Radioactive1 | Radioactive nuclei are unstable. They decay by emitting alpha or beta particles or gamma rays. |
| Isotopes1 | Atoms with the same atomic number, but different mass numbers are called isotopes. |
| Tracer1 | A small amount of radioactive isotope introduced into a system in order to follow the behavior of some component of that system is a tracer. |
| Impulse1 | Product of force and time interval over which it acts is the impulse. |
| Proof1 | A measure of ethanol concentration of an alcoholic beverage; proof is double the concentration by volume; for example, 50 percent by volume is 100 proof. |
| Domain1 | The domain is the set of values that the independent variable of a function may take on. A domain may be finite as in the set of numbers {1, 2, 3.n} or infinite as in all the numbers between 0 and 1. |
| Set1 | In mathematics a set is a collection of related objects. The mathematical usage is similar to the ordinary English meaning of the word. The objects that make up a set are called the elements of the set. If a set contains an unlimited number of elements it is an infinite set. Otherwise it is a finite set. |

## Chapter 10. Mass Balances and Hydraulic Flow Regimes

| | |
|---|---|
| Quantity1 | A numerical value either scalar or vector, which describes some attribute of an object like its position or its velocity . We sometimes speak of physical quantities to signify that we are talking about an object's properties or attributes as opposed to a purely mathematical quantity. |
| Density1 | The mass of a given volume of substance. It has units of kg/m3 or g/cm3. When the density is high the particles are closely packed. |
| Meter1 | Meter refers to SI unit of length. |
| Weight1 | Force of gravity of an object is called weight. |
| Activity1 | Number of decays per second of a radioactive substance is an activity. |
| Pulse1 | A wave of short duration confined to a small portion of the medium at any given time is called a pulse. A pulse is also called a wave pulse. |
| Range1 | The range is the set of values that the dependent variable of a function may take on. A range may be finite as in the set of numbers {1, 2, 3.n} or infinite as in all the mumbers between 0 and 1. |
| Secondary1 | The output coil of a transformer is referred to as secondary. |
| Resultant1 | Vector sum of two or more vectors is referred to as resultant. |
| Solids1 | A phase of matter with molecules that remain close to fixed equilibrium positions due to strong interactions between the molecules, resulting in the characteristic definite shape and definite volume are solids. |
| Color1 | The visual perception of light that enables human eyes to differentiate between wavelengths of the visible spectrum, with the longest wavelengths appearing red and the shortest appearing blue or violet is called color. |
| Solid1 | State of matter with fixed volume and shape is referred to as solid. |
| Vacuum1 | Vacuum refers to a region of space containing no matter. In practice, a region of gas at very low pressure. |

## Chapter 10. Mass Balances and Hydraulic Flow Regimes

| | |
|---|---|
| Waste1 | High-level waste is highly radioactive material arising from nuclear fission. It is recovered from reprocessing spent fuel, though some countries regard spent fuel itself as HLW and plan to dispose of it in that form. It requires very careful handling, storage and disposal. |
| Dose1 | More specifically referred to as 'absorbed dose', this is a measure of the energy deposited within a given mass of a patient. Absorbed dose is quantified by the unit called the 'rad'. |
| Gravity1 | A force with infinite range which acts between objects, such as planets, according to their mass is called gravity. |
| Node1 | Point where disturbances caused by two or more waves result in no displacement is the node. |
| Iteration1 | Iteration is the process of taking the value of the dependent variable of a function and feeding it back into the function as the independent variable. |
| Accuracy1 | Accuracy refers to closeness of a measurement to the standard value of that quantity. |
| Event1 | An event occurs when two particles collide or a single particle decay . Particle theories predict the probabilities of various events occurring when many similar collisions or decays are studied. They cannot predict the outcome for a single collision or decay. |
| Proportion1 | Proportion refers to two quantities are directly proportional if doubling one of them has the effect of doubling the other. On a graph we get a straight line through the origin. |
| Dose rate1 | Dose rate refers to a measure of the dose delivered per unit time. |

## Chapter 11. Screening and Sedimentation

| | |
|---|---|
| Physics1 | Study of matter and energy and their relationship is physics. |
| Fluid1 | Fluid refers to material that flows, i.e. liquids, gases, and plasmas. |
| Mechanics1 | The study of objects in motion. Mechanics is normally limited to a small number of large slow objects, as opposed to statistical mechanics which deals with large numbers of objects, relativistic mechanics which deals with objects moving near the speed of light and quantum mechanics which deals with objects more or less the size of atoms. Mechanics encompasses the topics of kinematics and dynamics . |
| Solids1 | A phase of matter with molecules that remain close to fixed equilibrium positions due to strong interactions between the molecules, resulting in the characteristic definite shape and definite volume are solids. |
| Matter1 | We call the commonly observed particles such as protons, neutrons and electrons matter particles, and their antiparticles, antimatter. |
| Velocity1 | The ratio of change in position with respect to the time interval over which the change occurred is referred to as velocity. |
| Function1 | A mathematical function is a rule relating two sets of objects. Here we will restrict ourselves to objects that are numbers or vectors. One of the sets is called the domain of the function, the other is called the range of the function. Functions are frequently expressed as equations as for example $Y=X+2$. This function is interpreted as follows. For every X in the domain, add 2 to it to get the corresponding Y in the range. Because we are free to choose any X we want, X is called the independent variable. Because once X is chosen Y is fixed, we call Y the dependent variable. |
| Bernoulli's equation1 | In an irrotational fluid, the sum of the static pressure, the weight of the fluid per unit mass times the height, and half the density times the velocity squared is constant throughout the fluid is called Bernoulli's equation. |
| Equation1 | An equation is a mathematical expression with an equal sign in it. It signifies that the numerical or vector value on one side of the = is the same as the numerical or vector value on the other side. An equation may include variables and parameters. If any of the variables are rates of change, the equation is called a differential equation. |
| Acceleration1 | Change in velocity divided by time interval over which it occurred is an acceleration. |

## Chapter 11. Screening and Sedimentation

| | |
|---|---|
| Gravity1 | A force with infinite range which acts between objects, such as planets, according to their mass is called gravity. |
| Secondary1 | The output coil of a transformer is referred to as secondary. |
| Range1 | The range is the set of values that the dependent variable of a function may take on. A range may be finite as in the set of numbers {1, 2, 3.n} or infinite as in all the mumbers between 0 and 1. |
| System1 | Defined collection of objects is called a system. |
| Density1 | The mass of a given volume of substance. It has units of kg/m3 or g/cm3. When the density is high the particles are closely packed. |
| Pressure1 | Force per unit area is referred to as the pressure. |
| Ultraviolet1 | Ultraviolet refers to electromagnetic waves with a wavelength shorter than that of visible light. |
| Light1 | Electromagnetic radiation with wavelengths between 400 and 700 nm that is visible is called light. |
| Trough1 | The point of maximum negative displacement on a transverse wave is called a trough. |
| Hydraulic1 | Hydraulic refers to using a fluid as a method of transmitting pressure. It allows forces to be magnified. |
| Speed1 | Speed refers to ratio of distance traveled to time interval. |
| Particle1 | In 'particle physics', a subatomic object with definite mass and charge . |
| Viscous fluid1 | Viscous fluid refers to fluid that creates force that opposes motion of objects through it. The force is proportional to object's speed. |
| Gravitational force1 | Attraction between two objects due to their mass is referred to as gravitational force. |
| Force1 | Force refers to agent that results in accelerating or deforming an object. |
| Resistance1 | Ratio of potential difference across device to current through it are called the resistance. |

## Chapter 11. Screening and Sedimentation

| | |
|---|---|
| Buoyant force1 | Upward force on an object immersed in fluid is called buoyant force. |
| Dimensional analysis1 | Checking a derived equation by making sure dimensions are the same on both sides is referred to as dimensional analysis. |
| Vector1 | Any quantity that has both magnitude and direction. Velocity is a vector. |
| Longitudinal1 | A type of wave motion in which the oscillations are parallel to the direction of wave travel is called longitudinal. |
| Energy1 | Energy refers to non-material property capable of causing changes in matter. |
| Turbulence1 | Unstable and disorderly motion, as when a smooth, flowing stream becomes a churning rapid is referred to as turbulence. |
| Dispersion1 | The splitting of light into its constituent colours is called dispersion. |
| Period1 | Period refers to time needed to repeat one complete cycle of motion. |
| Liquid1 | Material that has fixed volume but whose shape depends on the container is called liquid. |
| Movement1 | Change of position is called movement. |
| Efficiency1 | Ratio of output work to input work is called efficiency. |
| Distance1 | Distance refers to separation between two points. A scalar quantity. |
| Weight1 | Force of gravity of an object is called weight. |
| Frequency1 | Frequency refers to number of occurrences per unit time. |
| Quantity1 | A numerical value either scalar or vector, which describes some attribute of an object like its position or its velocity . We sometimes speak of physical quantities to signify that we are talking about an object's properties or attributes as opposed to a purely mathematical quantity. |
| Temperature1 | Temperature refers to measure of hotness of object on a quantitative scale. In gases, proportional to average kinetic energy of molecules. |

## Chapter 11. Screening and Sedimentation

| | |
|---|---|
| Second1 | Second is a SI unit of time. |
| Viscosity1 | A measure of how easy or otherwise a liquid can be poured. A high viscosity liquid flows very slowly. |
| Nominal1 | A nominal dimension is one that gives the intended or approximate size but this may (and often does) vary from the actual dimension. For example, a common lumber shape is a 2x4 but this is a nominal size and the actual dimensions are 1.5" x 3.5". The word is from Latin, of a name, nomin-, nomen name thus can be thought of as 'what we call it'. |
| Trajectory1 | The path followed by a projectile is referred to as trajectory. |
| Tangent1 | The ratio of the opposite side and the adjacent side is referred to as tangent. |
| Plastic1 | The behaviour of a material is plastic if it does not regain its original shape when the deforming force is removed. |
| Object1 | Object refers to source of diverging light rays; either luminous or illuminated. |
| Units1 | The units one uses should be of a size that makes sense for the particular subject at hand. It is easiest to define units in each area of science and then relate them to one another than to go around measuring particle masses in grams or cheese in proton mass units. In particle physics the standard unit is the unit of energy gev. One ev is the amount of energy that an electron gains when it moves through a potential difference of 1 Volt . G stands for Giga, or $10^9$. Thus a gev is a billion electron Volts. The mass-energy of a proton or neutron is approximately 1 gev. |
| Liquids1 | Liquids refers to a phase of matter composed of molecules that have interactions stronger than those found in a gas but not strong enough to keep the molecules near the equilibrium positions of a solid, resulting in the characteristic definite volume but indefinite shape. |
| Resultant1 | Vector sum of two or more vectors is referred to as resultant. |
| Plates1 | Large sections of the Earth's crust which float on the mantle are called plates. |
| Waste1 | High-level waste is highly radioactive material arising from nuclear fission. It is recovered from reprocessing spent fuel, though some countries regard spent fuel itself as HLW and plan to dispose of it in that form. It requires very careful handling, storage and disposal. |

## Chapter 11. Screening and Sedimentation

| | |
|---|---|
| Normal1 | Perpendicular to plane of interest is called normal. |
| State1 | Dynamical systems evolve over the course of time. The state of the system at any instant may be identified by the values of certain variables at that instant. For example specifying the angle from the vertical and the velocity of a frictionless pendulum allows us to predict its position and velocity at any future time. Therefore the state of the pendulum at any instant is its position and velocity. In this example the position and velocity are known as state variables. |
| Slope1 | Ratio of the vertical separation, or rise to the horizontal separation, or run are called the slope. |
| Flux1 | The flow of fluid, particles, or energy through a given area within a certain time. In astronomy, this term is often used to describe the rate at which light flows. For example, the amount of light striking a single square centimeter of a detector in one second is its flux. |
| Negative1 | The sign of the electric charge on the electron is negative. |
| Jet1 | The name physicists give to a cluster of particles emerging from a collision or decay event all traveling in roughly the same direction and carrying a significant fraction of the energy in the event. The particles in the jet are chiefly hadrons . |
| Gas1 | State of matter that expands to fill container is referred to as gas. |
| Decay1 | Any process in which a particle disappears and in its place two or more different particles appear is decay. |
| Gases1 | A phase of matter composed of molecules that are relatively far apart moving freely in a constant, random motion and have weak cohesive forces acting between them, resulting in the characteristic indefinite shape and indefinite volume of a gas are called the gases. |
| Set1 | In mathematics a set is a collection of related objects. The mathematical usage is similar to the ordinary English meaning of the word. The objects that make up a set are called the elements of the set. If a set contains an unlimited number of elements it is an infinite set. Otherwise it is a finite set. |
| Origin1 | Origin refers to the point in a reference frame from which measurements are made. It is the location of the zero value for each axis in the frame. |
| Axis1 | Axis refers to the imaginary line about which a planet or other object rotates . |

## Chapter 11. Screening and Sedimentation

| | |
|---|---|
| Momentum1 | Product of object's mass and velocity is referred to as momentum. |
| Friction1 | Force opposing relative motion of two objects are in contact is friction. |
| Boundary1 | Boundary refers to the division between two regions of differing physical properties . |
| Free fall1 | The motion of a body towards the earth when no other force except the force of gravity acts on it is called free fall. All freely falling bodies are weightless. |
| Crest1 | The point of maximum positive displacement on a transverse wave is called a crest. |
| Constant velocity1 | Velocity that does not change in time is called constant velocity. |
| Reference level1 | Location at which potential energy is chosen to be zero is a reference level. |
| Power1 | Power refers to rate of doing work; rate of energy conversion. |
| Centrifugal force1 | Centrifugal force refers to an apparent outward force on an object following a circular path that. This force is a consequence of the third law of motion . |
| Kinetic energy1 | Energy of object due to its motion is kinetic energy. |
| Random1 | With no set order or pattern, we have random. |
| Heat1 | Heat refers to quantity of energy transferred from one object to another because of a difference in temperature. |
| Iteration1 | Iteration is the process of taking the value of the dependent variable of a function and feeding it back into the function as the independent variable. |
| Time period1 | The time taken by a wave to travel through a distance equal to its wavelength is called its time period. |
| Current1 | Current refers to a flow of charge. Measured in amps. |
| Tracer1 | A small amount of radioactive isotope introduced into a system in order to follow the behavior of some component of that system is a tracer. |

## Chapter 12. Mass Transfer and Aeration

| | |
|---|---|
| Gases1 | A phase of matter composed of molecules that are relatively far apart moving freely in a constant, random motion and have weak cohesive forces acting between them, resulting in the characteristic indefinite shape and indefinite volume of a gas are called the gases. |
| Diffusion1 | The spreading out of a substance, due to the kinetic energy of its particles, to fill all of the available space are called diffusion. |
| Hydraulic1 | Hydraulic refers to using a fluid as a method of transmitting pressure. It allows forces to be magnified. |
| Dispersion1 | The splitting of light into its constituent colours is called dispersion. |
| Equilibrium1 | Equilibrium refers to condition in which net force is equal to zero. Condition in which net torque on object is zero. |
| Quantity1 | A numerical value either scalar or vector, which describes some attribute of an object like its position or its velocity . We sometimes speak of physical quantities to signify that we are talking about an object's properties or attributes as opposed to a purely mathematical quantity. |
| Random movement1 | Particles moving in no pattern or fixed way are called random movement. |
| Movement1 | Change of position is called movement. |
| Energy1 | Energy refers to non-material property capable of causing changes in matter. |
| Fluid1 | Fluid refers to material that flows, i.e. liquids, gases, and plasmas. |
| Liquid1 | Material that has fixed volume but whose shape depends on the container is called liquid. |
| Element1 | A pure substance that cannot be split up into anything simpler is called an element. |
| Gradient1 | Gradient refers to the slope of a graph. |
| Negative1 | The sign of the electric charge on the electron is negative. |

## Chapter 12. Mass Transfer and Aeration

| | |
|---|---|
| Equation1 | An equation is a mathematical expression with an equal sign in it. It signifies that the numerical or vector value on one side of the = is the same as the numerical or vector value on the other side. An equation may include variables and parameters. If any of the variables are rates of change, the equation is called a differential equation. |
| Distance1 | Distance refers to separation between two points. A scalar quantity. |
| Flux1 | The flow of fluid, particles, or energy through a given area within a certain time. In astronomy, this term is often used to describe the rate at which light flows. For example, the amount of light striking a single square centimeter of a detector in one second is its flux. |
| Velocity1 | The ratio of change in position with respect to the time interval over which the change occurred is referred to as velocity. |
| Longitudinal1 | A type of wave motion in which the oscillations are parallel to the direction of wave travel is called longitudinal. |
| Temperature1 | Temperature refers to measure of hotness of object on a quantitative scale. In gases, proportional to average kinetic energy of molecules. |
| Density1 | The mass of a given volume of substance. It has units of kg/m3 or g/cm3. When the density is high the particles are closely packed. |
| Gas1 | State of matter that expands to fill container is referred to as gas. |
| Kinetic energy1 | Energy of object due to its motion is kinetic energy. |
| Brownian motion1 | Brownian motion refers to the continuous random motion of solid microscopic particles when suspended in a fluid medium due to the consequence of ongoing bombardment by atoms and molecules. |
| Turbulence1 | Unstable and disorderly motion, as when a smooth, flowing stream becomes a churning rapid is referred to as turbulence. |
| Phase1 | The particles in a wave, which are in the same state of vibration, i.e., the same position and the same direction of motion are said to be in the same phase. |
| Pressure1 | Force per unit area is referred to as the pressure. |

## Chapter 12. Mass Transfer and Aeration

| | |
|---|---|
| State1 | Dynamical systems evolve over the course of time. The state of the system at any instant may be identified by the values of certain variables at that instant. For example specifying the angle from the vertical and the velocity of a frictionless pendulum allows us to predict its position and velocity at any future time. Therefore the state of the pendulum at any instant is its position and velocity. In this example the position and velocity are known as state variables. |
| Weight1 | Force of gravity of an object is called weight. |
| Function1 | A mathematical function is a rule relating two sets of objects. Here we will restrict ourselves to objects that are numbers or vectors. One of the sets is called the domain of the function, the other is called the range of the function. Functions are frequently expressed as equations as for example Y=X+2. This function is interpreted as follows. For every X in the domain, add 2 to it to get the corresponding Y in the range. Because we are free to choose any X we want, X is called the independent variable. Because once X is chosen Y is fixed, we call Y the dependent variable. |
| Efficiency1 | Ratio of output work to input work is called efficiency. |
| Period1 | Period refers to time needed to repeat one complete cycle of motion. |
| Slope1 | Ratio of the vertical separation, or rise to the horizontal separation, or run are called the slope. |
| Turbine1 | A rotary device that usually powers an electrical generator. The turbine may be turned by water, wind or high pressure steam. |
| Power1 | Power refers to rate of doing work; rate of energy conversion. |
| Ionized1 | An atom or a particle that has a net charge because it has gained or lost electrons is ionized. |
| System1 | Defined collection of objects is called a system. |
| Groundwater1 | Groundwater refers to water found in the voids or free space of soils and rocks underground. |
| Conversion1 | Chemical process turning U308 into UF6 preparatory to enrichment is conversion. |
| Explosion1 | Explosion refers to a very rapid reaction accompanied by a large expansion of gases. |
| Matter1 | We call the commonly observed particles such as protons, neutrons and electrons matter particles, and their antiparticles, antimatter. |

## Chapter 12. Mass Transfer and Aeration

| | |
|---|---|
| Normal1 | Perpendicular to plane of interest is called normal. |
| Displacement1 | Displacement refers to change in position. A vector quantity. |
| Property1 | A characteristic that is inherently associated with the object which is said to have that property. For example the mass of an object is one of its properties. So also might be color, density and many other characteristics. Properties are classified as extensive or intensive. Extensive properties increase in proportion to the size of the object, as mass does for example. Intensive properties are independent of the size of the object. The density for examples remains the same if I cut an object in half and throw half of it away. Things like an object's position or velocity are not considered to be properties of the object. They are not a characteristic of the object only but are also dependent on the reference frame in which the object is located. |
| Force1 | Force refers to agent that results in accelerating or deforming an object. |
| Gravity1 | A force with infinite range which acts between objects, such as planets,according to their mass is called gravity. |
| Plates1 | Large sections of the Earth's crust which float on the mantle are called plates. |
| Mechanics1 | The study of objects in motion. Mechanics is normally limited to a small number of large slow objects, as opposed to statistical mechanics which deals with large numbers of objects, relativistic mechanics which deals with objects moving near the speed of light and quantum mechanics which deals with objects more or less the size of atoms. Mechanics encompasses the topics of kinematics and dynamics . |
| Free fall1 | The motion of a body towards the earth when no other force except the force of gravity acts on it is called free fall. All freely falling bodies are weightless. |
| Acceleration1 | Change in velocity divided by time interval over which it occurred is an acceleration. |
| Range1 | The range is the set of values that the dependent variable of a function may take on. A range may be finite as in the set of numbers {1, 2, 3.n} or infinite as in all the mumbers between 0 and 1. |
| Trajectory1 | The path followed by a projectile is referred to as trajectory. |
| Initial velocity1 | Velocity of object at time t=0 is an initial velocity. |

CᴸᵃᵐΙΟΙ

## Chapter 12. Mass Transfer and Aeration

| | |
|---|---|
| Jet1 | The name physicists give to a cluster of particles emerging from a collision or decay event all traveling in roughly the same direction and carrying a significant fraction of the energy in the event. The particles in the jet are chiefly hadrons . |
| Sine1 | Sine refers to the ratio of the opposite side and the hypotenuse. |
| Particle1 | In 'particle physics', a subatomic object with definite mass and charge . |
| Nominal1 | A nominal dimension is one that gives the intended or approximate size but this may (and often does) vary from the actual dimension. For example, a common lumber shape is a 2x4 but this is a nominal size and the actual dimensions are 1.5" x 3.5". The word is from Latin, of a name, nomin-, nomen name thus can be thought of as 'what we call it'. |
| Terminal velocity1 | Velocity of falling object reached when force of air resistance equals weight is a terminal velocity. |
| Viscosity1 | A measure of how easy or otherwise a liquid can be poured. A high viscosity liquid flows very slowly. |
| Waste1 | High-level waste is highly radioactive material arising from nuclear fission. It is recovered from reprocessing spent fuel, though some countries regard spent fuel itself as HLW and plan to dispose of it in that form. It requires very careful handling, storage and disposal. |
| Dose1 | More specifically referred to as 'absorbed dose', this is a measure of the energy deposited within a given mass of a patient. Absorbed dose is quantified by the unit called the 'rad'. |
| Ion1 | Ion refers to a positively or negatively charged particle formed when an atom or group of atoms loses or gains electrons. |
| Interference1 | Phenomenon of light where the relative phase difference between two light waves produces light or dark spots, a result of light's wavelike nature is called the interference. |
| Set1 | In mathematics a set is a collection of related objects. The mathematical usage is similar to the ordinary English meaning of the word. The objects that make up a set are called the elements of the set. If a set contains an unlimited number of elements it is an infinite set. Otherwise it is a finite set. |

## Chapter 12. Mass Transfer and Aeration

| Units1 | The units one uses should be of a size that makes sense for the particular subject at hand. It is easiest to define units in each area of science and then relate them to one another than to go around measuring particle masses in grams or cheese in proton mass units. In particle physics the standard unit is the unit of energy gev. One ev is the amount of energy that an electron gains when it moves through a potential difference of 1 Volt . G stands for Giga, or $10^9$· Thus a gev is a billion electron Volts. The mass-energy of a proton or neutron is approximately 1 gev. |
|---|---|
| Plastic1 | The behaviour of a material is plastic if it does not regain its original shape when the deforming force is removed. |
| Barrier1 | Radiation-absorbing material, such as lead or concrete, used to reduce radiation exposure. A primary barrier attenuates useful beam to the required degree. A secondary barrier attenuates stray radiation to the required degree. |
| Solids1 | A phase of matter with molecules that remain close to fixed equilibrium positions due to strong interactions between the molecules, resulting in the characteristic definite shape and definite volume are solids. |
| Mole1 | The amount of substance containing $6 \times 10^{23}$ particles is the mole. |
| Speed1 | Speed refers to ratio of distance traveled to time interval. |
| Uniform velocity1 | When a body travels along a straight line in particular direction and covers equal distances in equal intervals of time it is said to have uniform velocity, we have uniform velocity. |
| Absorption1 | The transfer of energy to a medium, such as body tissues, as a radiation beam passes through the medium is absorption. |
| Liquids1 | Liquids refers to a phase of matter composed of molecules that have interactions stronger than those found in a gas but not strong enough to keep the molecules near the equilibrium positions of a solid, resulting in the characteristic definite volume but indefinite shape. |
| Physics1 | Study of matter and energy and their relationship is physics. |

## Chapter 13. Coagulation and Flocculation

| | |
|---|---|
| Period1 | Period refers to time needed to repeat one complete cycle of motion. |
| Solids1 | A phase of matter with molecules that remain close to fixed equilibrium positions due to strong interactions between the molecules, resulting in the characteristic definite shape and definite volume are solids. |
| Matter1 | We call the commonly observed particles such as protons, neutrons and electrons matter particles, and their antiparticles, antimatter. |
| Charge1 | A property of atomic particles. Electrons and protons have opposite charges and attract each other. The charge on an electron is negative and that on a proton is positive. Measured in coulombs. |
| Power1 | Power refers to rate of doing work; rate of energy conversion. |
| Magnitude1 | Magnitude refers to the size of a thing, without regard for its sign or direction. Similar to the absolute value of a number but applies to vectors as well. |
| Ion1 | Ion refers to a positively or negatively charged particle formed when an atom or group of atoms loses or gains electrons. |
| Charged1 | Object that has an unbalance of positive and negative electrical charges is referred to as charged. |
| Range1 | The range is the set of values that the dependent variable of a function may take on. A range may be finite as in the set of numbers {1, 2, 3.n} or infinite as in all the mumbers between 0 and 1. |
| Electrolyte1 | Electrolyte refers to water solution of ionic substances that conducts an electric current . |
| Negative1 | The sign of the electric charge on the electron is negative. |
| Temperature1 | Temperature refers to measure of hotness of object on a quantitative scale. In gases, proportional to average kinetic energy of molecules. |
| Efficiency1 | Ratio of output work to input work is called efficiency. |
| Metal1 | Matter having the physical properties of conductivity, malleability, ductility, and luster is referred to as metal. |

## Chapter 13. Coagulation and Flocculation

| | |
|---|---|
| Equilibrium1 | Equilibrium refers to condition in which net force is equal to zero. Condition in which net torque on object is zero. |
| Units1 | The units one uses should be of a size that makes sense for the particular subject at hand. It is easiest to define units in each area of science and then relate them to one another than to go around measuring particle masses in grams or cheese in proton mass units. In particle physics the standard unit is the unit of energy gev. One ev is the amount of energy that an electron gains when it moves through a potential difference of 1 Volt . G stands for Giga, or $10^9$. Thus a gev is a billion electron Volts. The mass-energy of a proton or neutron is approximately 1 gev. |
| Mole1 | The amount of substance containing $6 \times 10^{23}$ particles is the mole. |
| Waste1 | High-level waste is highly radioactive material arising from nuclear fission. It is recovered from reprocessing spent fuel, though some countries regard spent fuel itself as HLW and plan to dispose of it in that form. It requires very careful handling, storage and disposal. |
| Absorption1 | The transfer of energy to a medium, such as body tissues, as a radiation beam passes through the medium is absorption. |
| Ionized1 | An atom or a particle that has a net charge because it has gained or lost electrons is ionized. |
| Unstable1 | Matter that is capable of undergoing spontaneous change, as in a radioactive nuclide or an excited nuclear system. An unstable particle is any elementary particle that spontaneously decays into other particles. |
| Particle1 | In 'particle physics', a subatomic object with definite mass and charge . |
| Weight1 | Force of gravity of an object is called weight. |
| Dose1 | More specifically referred to as 'absorbed dose', this is a measure of the energy deposited within a given mass of a patient. Absorbed dose is quantified by the unit called the 'rad'. |
| Machine1 | Device that changes force needed to do work, is the machine. |
| Cycle1 | In wave motion, one cycle is a trough and a crest for a transverse wave, or a compression and a rarefaction for a longitudinal wave. |
| Distance1 | Distance refers to separation between two points. A scalar quantity. |

## Chapter 13. Coagulation and Flocculation

| | |
|---|---|
| Speed1 | Speed refers to ratio of distance traveled to time interval. |
| Phase1 | The particles in a wave, which are in the same state of vibration, i.e., the same position and the same direction of motion are said to be in the same phase. |
| Liquid1 | Material that has fixed volume but whose shape depends on the container is called liquid. |
| Frequency1 | Frequency refers to number of occurrences per unit time. |
| Diffusion1 | The spreading out of a substance, due to the kinetic energy of its particles, to fill all of the available space are called diffusion. |
| Brownian motion1 | Brownian motion refers to the continuous random motion of solid microscopic particles when suspended in a fluid medium due to the consequence of ongoing bombardment by atoms and molecules. |
| Turbulence1 | Unstable and disorderly motion, as when a smooth, flowing stream becomes a churning rapid is referred to as turbulence. |
| Velocity1 | The ratio of change in position with respect to the time interval over which the change occurred is referred to as velocity. |
| Gradient1 | Gradient refers to the slope of a graph. |
| Energy1 | Energy refers to non-material property capable of causing changes in matter. |
| Fluid1 | Fluid refers to material that flows, i.e. liquids, gases, and plasmas. |
| Force1 | Force refers to agent that results in accelerating or deforming an object. |
| Viscosity1 | A measure of how easy or otherwise a liquid can be poured. A high viscosity liquid flows very slowly. |
| Momentum1 | Product of object's mass and velocity is referred to as momentum. |
| Pressure1 | Force per unit area is referred to as the pressure. |
| Gravity1 | A force with infinite range which acts between objects, such as planets,according to their mass is called gravity. |

## Chapter 13. Coagulation and Flocculation

| | |
|---|---|
| Element1 | A pure substance that cannot be split up into anything simpler is called an element. |
| Equation1 | An equation is a mathematical expression with an equal sign in it. It signifies that the numerical or vector value on one side of the = is the same as the numerical or vector value on the other side. An equation may include variables and parameters. If any of the variables are rates of change, the equation is called a differential equation. |
| Hydraulic1 | Hydraulic refers to using a fluid as a method of transmitting pressure. It allows forces to be magnified. |
| Density1 | The mass of a given volume of substance. It has units of kg/m3 or g/cm3. When the density is high the particles are closely packed. |
| Acceleration1 | Change in velocity divided by time interval over which it occurred is an acceleration. |
| Supersaturated1 | Containing more than the normal saturation amount of a solute at a given temperature is supersaturated. |
| Interaction1 | A process in which a particle decays or it responds to a force due to the presence of another particle is called interaction. |
| Second1 | Second is a SI unit of time. |
| Function1 | A mathematical function is a rule relating two sets of objects. Here we will restrict ourselves to objects that are numbers or vectors. One of the sets is called the domain of the function, the other is called the range of the function. Functions are frequently expressed as equations as for example Y=X+2. This function is interpreted as follows. For every X in the domain, add 2 to it to get the corresponding Y in the range. Because we are free to choose any X we want, X is called the independent variable. Because once X is chosen Y is fixed, we call Y the dependent variable. |
| Gray1 | Gray refers to the derived SI unit of absorbed dose, defined as the absorbed dose in which the energy per unit mass imparted to the matter by ionizing radiation is 1 J/kg; it thus has units of J/kg. |
| Revolution1 | Revolution refers to the orbital motion of one object around another. The Earth revolves around the Sun in one year. The moon revolves around the Earth in approximately 28 days. |
| Friction1 | Force opposing relative motion of two objects are in contact is friction. |

137

## Chapter 13. Coagulation and Flocculation

| | |
|---|---|
| Kinetic energy1 | Energy of object due to its motion is kinetic energy. |
| Turbine1 | A rotary device that usually powers an electrical generator. The turbine may be turned by water, wind or high pressure steam. |
| Axis1 | Axis refers to the imaginary line about which a planet or other object rotates . |
| Liquids1 | Liquids refers to a phase of matter composed of molecules that have interactions stronger than those found in a gas but not strong enough to keep the molecules near the equilibrium positions of a solid, resulting in the characteristic definite volume but indefinite shape. |
| Gas1 | State of matter that expands to fill container is referred to as gas. |
| Mixture1 | Matter made of unlike parts that have a variable composition and can be separated into their component parts by physical means is a mixture. |
| Boyle's law1 | The product of the pressure and the volume of an ideal gas at constant temperature is a constant is Boyle's law. |
| Compression1 | In a sound wave, a compression is a region in the material that transmits the wave where the particles are closer together than normal. |
| Electrical energy1 | Electrical energy refers to a form of energy from electromagnetic interactions; one of five forms of energy-mechanical, chemical, radiant, electrical, and nuclear . |
| Dispersion1 | The splitting of light into its constituent colours is called dispersion. |
| Jet1 | The name physicists give to a cluster of particles emerging from a collision or decay event all traveling in roughly the same direction and carrying a significant fraction of the energy in the event. The particles in the jet are chiefly hadrons . |
| Bernoulli's equation1 | In an irrotational fluid, the sum of the static pressure, the weight of the fluid per unit mass times the height, and half the density times the velocity squared is constant throughout the fluid is called Bernoulli's equation. |
| Vector1 | Any quantity that has both magnitude and direction. Velocity is a vector. |
| Normal1 | Perpendicular to plane of interest is called normal. |

## Chapter 13. Coagulation and Flocculation

| | |
|---|---|
| Mechanics1 | The study of objects in motion. Mechanics is normally limited to a small number of large slow objects, as opposed to statistical mechanics which deals with large numbers of objects, relativistic mechanics which deals with objects moving near the speed of light and quantum mechanics which deals with objects more or less the size of atoms. Mechanics encompasses the topics of kinematics and dynamics . |
| System1 | Defined collection of objects is called a system. |
| Movement1 | Change of position is called movement. |
| Set1 | In mathematics a set is a collection of related objects. The mathematical usage is similar to the ordinary English meaning of the word. The objects that make up a set are called the elements of the set. If a set contains an unlimited number of elements it is an infinite set. Otherwise it is a finite set. |
| Average velocity1 | Velocity measured over a finite time interval is referred to as average velocity. |
| Quantity1 | A numerical value either scalar or vector, which describes some attribute of an object like its position or its velocity . We sometimes speak of physical quantities to signify that we are talking about an object's properties or attributes as opposed to a purely mathematical quantity. |
| Secondary1 | The output coil of a transformer is referred to as secondary. |
| Nominal1 | A nominal dimension is one that gives the intended or approximate size but this may (and often does) vary from the actual dimension. For example, a common lumber shape is a 2x4 but this is a nominal size and the actual dimensions are 1.5" x 3.5". The word is from Latin, of a name, nomin-, nomen name thus can be thought of as 'what we call it'. |
| Intensity1 | The amount of radiation, for example, the number of photons arriving in a given time, is called intensity. |
| Dynamics1 | Study of motion of particles acted on by forces is called dynamics. |
| Fluids1 | Fluids refers to matter that has the ability to flow or be poured; the individual molecules of a fluid are able to move, rolling over or by one another . |

## Chapter 14. Filtration

| | |
|---|---|
| Solids1 | A phase of matter with molecules that remain close to fixed equilibrium positions due to strong interactions between the molecules, resulting in the characteristic definite shape and definite volume are solids. |
| Diffusion1 | The spreading out of a substance, due to the kinetic energy of its particles, to fill all of the available space are called diffusion. |
| Hydraulic1 | Hydraulic refers to using a fluid as a method of transmitting pressure. It allows forces to be magnified. |
| Nominal1 | A nominal dimension is one that gives the intended or approximate size but this may (and often does) vary from the actual dimension. For example, a common lumber shape is a 2x4 but this is a nominal size and the actual dimensions are 1.5" x 3.5". The word is from Latin, of a name, nomin-, nomen name thus can be thought of as 'what we call it'. |
| Meter1 | Meter refers to SI unit of length. |
| Inertia1 | Inertia refers to tendency of object not to change its motion. |
| Phase1 | The particles in a wave, which are in the same state of vibration, i.e., the same position and the same direction of motion are said to be in the same phase. |
| Period1 | Period refers to time needed to repeat one complete cycle of motion. |
| Cycle1 | In wave motion, one cycle is a trough and a crest for a transverse wave, or a compression and a rarefaction for a longitudinal wave. |
| Velocity1 | The ratio of change in position with respect to the time interval over which the change occurred is referred to as velocity. |
| Particle1 | In 'particle physics', a subatomic object with definite mass and charge . |
| Matter1 | We call the commonly observed particles such as protons, neutrons and electrons matter particles, and their antiparticles, antimatter. |
| Properties1 | Properties refers to qualities or attributes that, taken together, are usually unique to an object; for example, color, texture, and size. |
| Weight1 | Force of gravity of an object is called weight. |

## Chapter 14. Filtration

---

| | |
|---|---|
| Resistance1 | Ratio of potential difference across device to current through it are called the resistance. |
| Equation1 | An equation is a mathematical expression with an equal sign in it. It signifies that the numerical or vector value on one side of the = is the same as the numerical or vector value on the other side. An equation may include variables and parameters. If any of the variables are rates of change, the equation is called a differential equation. |
| Density1 | The mass of a given volume of substance. It has units of kg/m3 or g/cm3. When the density is high the particles are closely packed. |
| Plastic1 | The behaviour of a material is plastic if it does not regain its original shape when the deforming force is removed. |
| Proportion1 | Proportion refers to two quantities are directly proportional if doubling one of them has the effect of doubling the other. On a graph we get a straight line through the origin. |
| Constant velocity1 | Velocity that does not change in time is called constant velocity. |
| Force1 | Force refers to agent that results in accelerating or deforming an object. |
| Fluid1 | Fluid refers to material that flows, i.e. liquids, gases, and plasmas. |
| Function1 | A mathematical function is a rule relating two sets of objects. Here we will restrict ourselves to objects that are numbers or vectors. One of the sets is called the domain of the function, the other is called the range of the function. Functions are frequently expressed as equations as for example Y=X+2. This function is interpreted as follows. For every X in the domain, add 2 to it to get the corresponding Y in the range. Because we are free to choose any X we want, X is called the independent variable. Because once X is chosen Y is fixed, we call Y the dependent variable. |
| Pressure1 | Force per unit area is referred to as the pressure. |
| Position1 | Separation between object and a reference point is a position. |
| Friction1 | Force opposing relative motion of two objects are in contact is friction. |
| Average velocity1 | Velocity measured over a finite time interval is referred to as average velocity. |
| Second1 | Second is a SI unit of time. |

## Chapter 14. Filtration

| Temperature1 | Temperature refers to measure of hotness of object on a quantitative scale. In gases, proportional to average kinetic energy of molecules. |
| --- | --- |
| Units1 | The units one uses should be of a size that makes sense for the particular subject at hand. It is easiest to define units in each area of science and then relate them to one another than to go around measuring particle masses in grams or cheese in proton mass units. In particle physics the standard unit is the unit of energy gev. One ev is the amount of energy that an electron gains when it moves through a potential difference of 1 Volt . G stands for Giga, or $10^9$. Thus a gev is a billion electron Volts. The mass-energy of a proton or neutron is approximately 1 gev. |
| Set1 | In mathematics a set is a collection of related objects. The mathematical usage is similar to the ordinary English meaning of the word. The objects that make up a set are called the elements of the set. If a set contains an unlimited number of elements it is an infinite set. Otherwise it is a finite set. |
| Solid1 | State of matter with fixed volume and shape is referred to as solid. |
| System1 | Defined collection of objects is called a system. |
| Energy1 | Energy refers to non-material property capable of causing changes in matter. |
| Sound1 | A longitudinal wave, usually through the air but can travel through liquids and solids is called sound. |
| Physics1 | Study of matter and energy and their relationship is physics. |
| 2+b*x+c . | A function involving the second and lower power, and none higher, of the independent variable. A quadratic function may contain $x^2$ explicitly or it may contain terms like x*, where the second power of x is implied. In general a quadratic may be written as y=a*x |
| Movement1 | Change of position is called movement. |
| Slope1 | Ratio of the vertical separation, or rise to the horizontal separation, or run are called the slope. |
| Range1 | The range is the set of values that the dependent variable of a function may take on. A range may be finite as in the set of numbers {1, 2, 3.n} or infinite as in all the mumbers between 0 and 1. |

## Chapter 14. Filtration

| | |
|---|---|
| Viscosity1 | A measure of how easy or otherwise a liquid can be poured. A high viscosity liquid flows very slowly. |
| Secondary1 | The output coil of a transformer is referred to as secondary. |
| Waste1 | High-level waste is highly radioactive material arising from nuclear fission. It is recovered from reprocessing spent fuel, though some countries regard spent fuel itself as HLW and plan to dispose of it in that form. It requires very careful handling, storage and disposal. |
| Gas1 | State of matter that expands to fill container is referred to as gas. |
| Gases1 | A phase of matter composed of molecules that are relatively far apart moving freely in a constant, random motion and have weak cohesive forces acting between them, resulting in the characteristic indefinite shape and indefinite volume of a gas are called the gases. |
| Negative1 | The sign of the electric charge on the electron is negative. |
| Distance1 | Distance refers to separation between two points. A scalar quantity. |
| Bernoulli's equation1 | In an irrotational fluid, the sum of the static pressure, the weight of the fluid per unit mass times the height, and half the density times the velocity squared is constant throughout the fluid is called Bernoulli's equation. |
| Color1 | The visual perception of light that enables human eyes to differentiate between wavelengths of the visible spectrum, with the longest wavelengths appearing red and the shortest appearing blue or violet is called color. |
| Metal1 | Matter having the physical properties of conductivity, malleability, ductility, and luster is referred to as metal. |
| Activity1 | Number of decays per second of a radioactive substance is an activity. |
| Decay1 | Any process in which a particle disappears and in its place two or more different particles appear is decay. |
| Unstable1 | Matter that is capable of undergoing spontaneous change, as in a radioactive nuclide or an excited nuclear system. An unstable particle is any elementary particle that spontaneously decays into other particles. |

## Chapter 14. Filtration

| | |
|---|---|
| Normal1 | Perpendicular to plane of interest is called normal. |
| Time period1 | The time taken by a wave to travel through a distance equal to its wavelength is called its time period. |
| State1 | Dynamical systems evolve over the course of time. The state of the system at any instant may be identified by the values of certain variables at that instant. For example specifying the angle from the vertical and the velocity of a frictionless pendulum allows us to predict its position and velocity at any future time. Therefore the state of the pendulum at any instant is its position and velocity. In this example the position and velocity are known as state variables. |
| Stable1 | Does not decay is stable. |

## Chapter 15. Physical-Chemical Treatment for Dissolved Constituents

| | |
|---|---|
| Ion1 | Ion refers to a positively or negatively charged particle formed when an atom or group of atoms loses or gains electrons. |
| Color1 | The visual perception of light that enables human eyes to differentiate between wavelengths of the visible spectrum, with the longest wavelengths appearing red and the shortest appearing blue or violet is called color. |
| Neutral1 | Object that has no net electric charge is called neutral. |
| Weight1 | Force of gravity of an object is called weight. |
| Normal1 | Perpendicular to plane of interest is called normal. |
| Equilibrium1 | Equilibrium refers to condition in which net force is equal to zero. Condition in which net torque on object is zero. |
| Temperature1 | Temperature refers to measure of hotness of object on a quantitative scale. In gases, proportional to average kinetic energy of molecules. |
| Stable1 | Does not decay is stable. |
| Solids1 | A phase of matter with molecules that remain close to fixed equilibrium positions due to strong interactions between the molecules, resulting in the characteristic definite shape and definite volume are solids. |
| Second1 | Second is a SI unit of time. |
| Gas1 | State of matter that expands to fill container is referred to as gas. |
| Mole1 | The amount of substance containing $6 \times 10^{23}$ particles is the mole. |
| Groundwater1 | Groundwater refers to water found in the voids or free space of soils and rocks underground. |
| Dose1 | More specifically referred to as 'absorbed dose', this is a measure of the energy deposited within a given mass of a patient. Absorbed dose is quantified by the unit called the 'rad'. |
| Charge1 | A property of atomic particles. Electrons and protons have opposite charges and attract each other. The charge on an electron is negative and that on a proton is positive. Measured in coulombs. |

## Chapter 15. Physical-Chemical Treatment for Dissolved Constituents

| | |
|---|---|
| Negative1 | The sign of the electric charge on the electron is negative. |
| State1 | Dynamical systems evolve over the course of time. The state of the system at any instant may be identified by the values of certain variables at that instant. For example specifying the angle from the vertical and the velocity of a frictionless pendulum allows us to predict its position and velocity at any future time. Therefore the state of the pendulum at any instant is its position and velocity. In this example the position and velocity are known as state variables. |
| Quantity1 | A numerical value either scalar or vector, which describes some attribute of an object like its position or its velocity . We sometimes speak of physical quantities to signify that we are talking about an object's properties or attributes as opposed to a purely mathematical quantity. |
| System1 | Defined collection of objects is called a system. |
| Barrier1 | Radiation-absorbing material, such as lead or concrete, used to reduce radiation exposure. A primary barrier attenuates useful beam to the required degree. A secondary barrier attenuates stray radiation to the required degree. |
| Equation1 | An equation is a mathematical expression with an equal sign in it. It signifies that the numerical or vector value on one side of the = is the same as the numerical or vector value on the other side. An equation may include variables and parameters. If any of the variables are rates of change, the equation is called a differential equation. |
| Activity1 | Number of decays per second of a radioactive substance is an activity. |
| Meter1 | Meter refers to SI unit of length. |
| Range1 | The range is the set of values that the dependent variable of a function may take on. A range may be finite as in the set of numbers {1, 2, 3.n} or infinite as in all the mumbers between 0 and 1. |
| Crystalline1 | A regularly repeated crystal-like substructure is crystalline. |
| Metal1 | Matter having the physical properties of conductivity, malleability, ductility, and luster is referred to as metal. |
| Matter1 | We call the commonly observed particles such as protons, neutrons and electrons matter particles, and their antiparticles, antimatter. |

| | |
|---|---|
| Conversion1 | Chemical process turning U308 into UF6 preparatory to enrichment is conversion. |
| Units1 | The units one uses should be of a size that makes sense for the particular subject at hand. It is easiest to define units in each area of science and then relate them to one another than to go around measuring particle masses in grams or cheese in proton mass units. In particle physics the standard unit is the unit of energy gev. One ev is the amount of energy that an electron gains when it moves through a potential difference of 1 Volt . G stands for Giga, or $10^9$. Thus a gev is a billion electron Volts. The mass-energy of a proton or neutron is approximately 1 gev. |
| Pressure1 | Force per unit area is referred to as the pressure. |
| Frequency1 | Frequency refers to number of occurrences per unit time. |
| Waste1 | High-level waste is highly radioactive material arising from nuclear fission. It is recovered from reprocessing spent fuel, though some countries regard spent fuel itself as HLW and plan to dispose of it in that form. It requires very careful handling, storage and disposal. |
| Function1 | A mathematical function is a rule relating two sets of objects. Here we will restrict ourselves to objects that are numbers or vectors. One of the sets is called the domain of the function, the other is called the range of the function. Functions are frequently expressed as equations as for example Y=X+2. This function is interpreted as follows. For every X in the domain, add 2 to it to get the corresponding Y in the range. Because we are free to choose any X we want, X is called the independent variable. Because once X is chosen Y is fixed, we call Y the dependent variable. |
| Secondary1 | The output coil of a transformer is referred to as secondary. |
| Solid1 | State of matter with fixed volume and shape is referred to as solid. |
| Velocity1 | The ratio of change in position with respect to the time interval over which the change occurred is referred to as velocity. |
| Particle1 | In 'particle physics', a subatomic object with definite mass and charge . |
| Unstable1 | Matter that is capable of undergoing spontaneous change, as in a radioactive nuclide or an excited nuclear system. An unstable particle is any elementary particle that spontaneously decays into other particles. |

## Chapter 15. Physical-Chemical Treatment for Dissolved Constituents

| | |
|---|---|
| Force1 | Force refers to agent that results in accelerating or deforming an object. |
| Gradient1 | Gradient refers to the slope of a graph. |
| Efficiency1 | Ratio of output work to input work is called efficiency. |
| Energy1 | Energy refers to non-material property capable of causing changes in matter. |
| Fluid1 | Fluid refers to material that flows, i.e. liquids, gases, and plasmas. |
| Liquid1 | Material that has fixed volume but whose shape depends on the container is called liquid. |
| Cell1 | In electricity, a cell is a combination of metals and chemicals that produces a voltage and can cause a current. |
| Distance1 | Distance refers to separation between two points. A scalar quantity. |
| Flux1 | The flow of fluid, particles, or energy through a given area within a certain time. In astronomy, this term is often used to describe the rate at which light flows. For example, the amount of light striking a single square centimeter of a detector in one second is its flux. |
| Resistance1 | Ratio of potential difference across device to current through it are called the resistance. |
| Polarization1 | A polarized particle beam is a beam of particles whose spins are aligned in a particular direction. The polarization of the beam is the fraction of the particles with the desired alignment. |
| Movement1 | Change of position is called movement. |
| Diffusion1 | The spreading out of a substance, due to the kinetic energy of its particles, to fill all of the available space are called diffusion. |
| Voltage1 | Voltage refers to potential difference. It is a measure of the change in energy that one coulomb of electric charge undergoes when moved between 2 points. |
| Cathode1 | Cathode refers to a negatively charged electrode. |
| Boundary1 | Boundary refers to the division between two regions of differing physical properties . |

## Chapter 15. Physical-Chemical Treatment for Dissolved Constituents

| | |
|---|---|
| Micron1 | Micron refers to one millionth of a meter; also known as a micrometer. |
| Shield1 | A mass of attenuating material used to prevent or reduce the passage of radiation or particles is called shield. |
| Transverse1 | A type of wave in which the vibrations are at right angles to the direction of wave motion is referred to as the transverse. |
| Hydraulic1 | Hydraulic refers to using a fluid as a method of transmitting pressure. It allows forces to be magnified. |
| Liquids1 | Liquids refers to a phase of matter composed of molecules that have interactions stronger than those found in a gas but not strong enough to keep the molecules near the equilibrium positions of a solid, resulting in the characteristic definite volume but indefinite shape. |
| Mercury1 | Mercury refers to the innermost planet in the solar system, and a metallic element that is liquid at room temperature. |
| Gases1 | A phase of matter composed of molecules that are relatively far apart moving freely in a constant, random motion and have weak cohesive forces acting between them, resulting in the characteristic indefinite shape and indefinite volume of a gas are called the gases. |
| Density1 | The mass of a given volume of substance. It has units of kg/m3 or g/cm3. When the density is high the particles are closely packed. |
| Properties1 | Properties refers to qualities or attributes that, taken together, are usually unique to an object; for example, color, texture, and size. |
| Phase1 | The particles in a wave, which are in the same state of vibration, i.e., the same position and the same direction of motion are said to be in the same phase. |
| Proportionality constant1 | A constant applied to a proportionality statement that transforms the statement into an equation is a proportionality constant. |
| Period1 | Period refers to time needed to repeat one complete cycle of motion. |

## Chapter 15. Physical-Chemical Treatment for Dissolved Constituents

| | |
|---|---|
| Nominal1 | A nominal dimension is one that gives the intended or approximate size but this may (and often does) vary from the actual dimension. For example, a common lumber shape is a 2x4 but this is a nominal size and the actual dimensions are 1.5" x 3.5". The word is from Latin, of a name, nomin-, nomen name thus can be thought of as 'what we call it'. |
| Slope1 | Ratio of the vertical separation, or rise to the horizontal separation, or run are called the slope. |
| Origin1 | Origin refers to the point in a reference frame from which measurements are made. It is the location of the zero value for each axis in the frame. |
| Current1 | Current refers to a flow of charge. Measured in amps. |
| Set1 | In mathematics a set is a collection of related objects. The mathematical usage is similar to the ordinary English meaning of the word. The objects that make up a set are called the elements of the set. If a set contains an unlimited number of elements it is an infinite set. Otherwise it is a finite set. |
| Wave1 | A set of oscillations or vibrations that transfer energy without any transfer of mass is the wave. |
| Power1 | Power refers to rate of doing work; rate of energy conversion. |
| Atom1 | Atom refers to the smallest particle of an element which can exist. |

## Chapter 16. Disinfection

| Contamination1 | Radioactive material deposited or dispersed in materials or places where it is not wanted is referred to as contamination. |
| --- | --- |
| Efficiency1 | Ratio of output work to input work is called efficiency. |
| Energy1 | Energy refers to non-material property capable of causing changes in matter. |
| Ultraviolet1 | Ultraviolet refers to electromagnetic waves with a wavelength shorter than that of visible light. |
| System1 | Defined collection of objects is called a system. |
| Current1 | Current refers to a flow of charge. Measured in amps. |
| Function1 | A mathematical function is a rule relating two sets of objects. Here we will restrict ourselves to objects that are numbers or vectors. One of the sets is called the domain of the function, the other is called the range of the function. Functions are frequently expressed as equations as for example Y=X+2. This function is interpreted as follows. For every X in the domain, add 2 to it to get the corresponding Y in the range. Because we are free to choose any X we want, X is called the independent variable. Because once X is chosen Y is fixed, we call Y the dependent variable. |
| Dose1 | More specifically referred to as 'absorbed dose', this is a measure of the energy deposited within a given mass of a patient. Absorbed dose is quantified by the unit called the 'rad'. |
| Power1 | Power refers to rate of doing work; rate of energy conversion. |
| Units1 | The units one uses should be of a size that makes sense for the particular subject at hand. It is easiest to define units in each area of science and then relate them to one another than to go around measuring particle masses in grams or cheese in proton mass units. In particle physics the standard unit is the unit of energy gev. One ev is the amount of energy that an electron gains when it moves through a potential difference of 1 Volt . G stands for Giga, or $10^9$. Thus a gev is a billion electron Volts. The mass-energy of a proton or neutron is approximately 1 gev. |
| Gas1 | State of matter that expands to fill container is referred to as gas. |
| Equilibrium1 | Equilibrium refers to condition in which net force is equal to zero. Condition in which net torque on object is zero. |

## Chapter 16. Disinfection

| | |
|---|---|
| Second1 | Second is a SI unit of time. |
| Range1 | The range is the set of values that the dependent variable of a function may take on. A range may be finite as in the set of numbers {1, 2, 3.n} or infinite as in all the mumbers between 0 and 1. |
| Ion1 | Ion refers to a positively or negatively charged particle formed when an atom or group of atoms loses or gains electrons. |
| Temperature1 | Temperature refers to measure of hotness of object on a quantitative scale. In gases, proportional to average kinetic energy of molecules. |
| Equation1 | An equation is a mathematical expression with an equal sign in it. It signifies that the numerical or vector value on one side of the = is the same as the numerical or vector value on the other side. An equation may include variables and parameters. If any of the variables are rates of change, the equation is called a differential equation. |
| Matter1 | We call the commonly observed particles such as protons, neutrons and electrons matter particles, and their antiparticles, antimatter. |
| Electron1 | Subatomic particle of small mass and negative charge found in every atom is called the electron. |
| Molecule1 | The smallest part of an element or compound that can exist on its own is called the molecule. |
| Stable1 | Does not decay is stable. |
| Unstable1 | Matter that is capable of undergoing spontaneous change, as in a radioactive nuclide or an excited nuclear system. An unstable particle is any elementary particle that spontaneously decays into other particles. |
| Mole1 | The amount of substance containing $6 \times 10^{23}$ particles is the mole. |
| Secondary1 | The output coil of a transformer is referred to as secondary. |
| Decay1 | Any process in which a particle disappears and in its place two or more different particles appear is decay. |

## Chapter 16. Disinfection

| | |
|---|---|
| Turbulence1 | Unstable and disorderly motion, as when a smooth, flowing stream becomes a churning rapid is referred to as turbulence. |
| Evaporation1 | Change from liquid to vapor state are called the evaporation. |
| Wavelength1 | Distance between corresponding points on two successive waves is referred to as wavelength. |
| Voltage1 | Voltage refers to potential difference. It is a measure of the change in energy that one coulomb of electric charge undergoes when moved between 2 points. |
| Resistance1 | Ratio of potential difference across device to current through it are called the resistance. |
| Quantity1 | A numerical value either scalar or vector, which describes some attribute of an object like its position or its velocity . We sometimes speak of physical quantities to signify that we are talking about an object's properties or attributes as opposed to a purely mathematical quantity. |
| Activity1 | Number of decays per second of a radioactive substance is an activity. |
| Properties1 | Properties refers to qualities or attributes that, taken together, are usually unique to an object; for example, color, texture, and size. |
| Frequency1 | Frequency refers to number of occurrences per unit time. |
| Set1 | In mathematics a set is a collection of related objects. The mathematical usage is similar to the ordinary English meaning of the word. The objects that make up a set are called the elements of the set. If a set contains an unlimited number of elements it is an infinite set. Otherwise it is a finite set. |
| Color1 | The visual perception of light that enables human eyes to differentiate between wavelengths of the visible spectrum, with the longest wavelengths appearing red and the shortest appearing blue or violet is called color. |
| Pressure1 | Force per unit area is referred to as the pressure. |
| Dew1 | Dew refers to condensation of water vapor into droplets of liquid on surfaces . |
| Transformer1 | Device to transform energy from one electrical circuit to another by means of mutual inductance between two coils is referred to as a transformer. |

## Chapter 16. Disinfection

| | |
|---|---|
| Generator1 | Generator produces electricity when a magnet spins inside a coil of wire or a coil of wire spins inside a magnetic field. |
| Corona1 | The outermost layer of the Sun's atmosphere, visible to the eye during a total solar eclipse; it can also be observed through special filters and best of all, by X-ray cameras aboart satellites. The corona is very hot, up to 1-1.5 million degrees centigrade, and is the source of the solar wind. |
| Vapor1 | The gaseous state of a substance that is normally in the liquid state is vapor. |
| Alternating current1 | Alternating current refers to an electric current that changes direction periodically. |
| Metal1 | Matter having the physical properties of conductivity, malleability, ductility, and luster is referred to as metal. |
| Diffusion1 | The spreading out of a substance, due to the kinetic energy of its particles, to fill all of the available space are called diffusion. |
| Radiation1 | Electromagnetic waves that carry energy are referred to as radiation. |
| Plastic1 | The behaviour of a material is plastic if it does not regain its original shape when the deforming force is removed. |
| Ionizing radiation1 | Ionizing radiation refers to particles or waves that can remove electrons from atoms, molecules, or atoms in a solid. |
| Electromagnetic radiation1 | Energy carried by electromagnetic waves throughout space is electromagnetic radiation. |
| Absorption1 | The transfer of energy to a medium, such as body tissues, as a radiation beam passes through the medium is absorption. |
| Solids1 | A phase of matter with molecules that remain close to fixed equilibrium positions due to strong interactions between the molecules, resulting in the characteristic definite shape and definite volume are solids. |
| Shield1 | A mass of attenuating material used to prevent or reduce the passage of radiation or particles is called shield. |

## Chapter 16. Disinfection

| Light1 | Electromagnetic radiation with wavelengths between 400 and 700 nm that is visible is called light. |
| --- | --- |
| Lamp1 | A device that gives out light when an electric current passes through it is called a lamp. |
| Ionized1 | An atom or a particle that has a net charge because it has gained or lost electrons is ionized. |
| Mercury1 | Mercury refers to the innermost planet in the solar system, and a metallic element that is liquid at room temperature. |
| Intensity1 | The amount of radiation, for example, the number of photons arriving in a given time, is called intensity. |
| Dispersion1 | The splitting of light into its constituent colours is called dispersion. |
| Period1 | Period refers to time needed to repeat one complete cycle of motion. |
| Cell1 | In electricity, a cell is a combination of metals and chemicals that produces a voltage and can cause a current. |
| Radioactive1 | Radioactive nuclei are unstable. They decay by emitting alpha or beta particles or gamma rays. |
| Electron beam1 | The stream of electrons generated by the electron gun and accelerated by the accelerator guide is called the electron beam. |
| Beam1 | Beam refers to a unidirectional or approximately unidirectional flow of electromagnetic radiation or particles. |
| Strain1 | Strain is the change in length per unit length. It is normally computed as $(Lf - Lo) / Lo$ where $Lf$ is the final length and $Lo$ is the initial length. When testing materials, a gage length is normally specified known; this represents $Lo$. |
| Weight1 | Force of gravity of an object is called weight. |
| Sun1 | Sun refers to the star at the centre of the solar system. The star the Earth orbits. |

## Chapter 16. Disinfection

| | |
|---|---|
| Barrier1 | Radiation-absorbing material, such as lead or concrete, used to reduce radiation exposure. A primary barrier attenuates useful beam to the required degree. A secondary barrier attenuates stray radiation to the required degree. |
| Liquid1 | Material that has fixed volume but whose shape depends on the container is called liquid. |
| Infrared1 | A type of electromagnetic radiation with a wavelength longer than that of light is called infrared. |
| Normal1 | Perpendicular to plane of interest is called normal. |
| Dose equivalent1 | Dose equivalent refers to a parameter used to express the risk of the deleterious effects of ionization radiation upon living organisms. For radiation protection purposes, the quantity of the effective irradiation incurred by exposed persons, measured on a common scale in sievert or rem . |
| Phase1 | The particles in a wave, which are in the same state of vibration, i.e., the same position and the same direction of motion are said to be in the same phase. |
| Turbine1 | A rotary device that usually powers an electrical generator. The turbine may be turned by water, wind or high pressure steam. |
| State1 | Dynamical systems evolve over the course of time. The state of the system at any instant may be identified by the values of certain variables at that instant. For example specifying the angle from the vertical and the velocity of a frictionless pendulum allows us to predict its position and velocity at any future time. Therefore the state of the pendulum at any instant is its position and velocity. In this example the position and velocity are known as state variables. |
| Ionization1 | The process by which a neutral atom or molecule acquires a positive or negative charge is called ionization. |
| Tracer1 | A small amount of radioactive isotope introduced into a system in order to follow the behavior of some component of that system is a tracer. |

## Chapter 17. Aerobic Biological Treatment

| | |
|---|---|
| Matter1 | We call the commonly observed particles such as protons, neutrons and electrons matter particles, and their antiparticles, antimatter. |
| Decay1 | Any process in which a particle disappears and in its place two or more different particles appear is decay. |
| Waste1 | High-level waste is highly radioactive material arising from nuclear fission. It is recovered from reprocessing spent fuel, though some countries regard spent fuel itself as HLW and plan to dispose of it in that form. It requires very careful handling, storage and disposal. |
| Second1 | Second is a SI unit of time. |
| Solids1 | A phase of matter with molecules that remain close to fixed equilibrium positions due to strong interactions between the molecules, resulting in the characteristic definite shape and definite volume are solids. |
| Period1 | Period refers to time needed to repeat one complete cycle of motion. |
| Energy1 | Energy refers to non-material property capable of causing changes in matter. |
| Quantity1 | A numerical value either scalar or vector, which describes some attribute of an object like its position or its velocity . We sometimes speak of physical quantities to signify that we are talking about an object's properties or attributes as opposed to a purely mathematical quantity. |
| System1 | Defined collection of objects is called a system. |
| Cycle1 | In wave motion, one cycle is a trough and a crest for a transverse wave, or a compression and a rarefaction for a longitudinal wave. |
| Equation1 | An equation is a mathematical expression with an equal sign in it. It signifies that the numerical or vector value on one side of the = is the same as the numerical or vector value on the other side. An equation may include variables and parameters. If any of the variables are rates of change, the equation is called a differential equation. |
| Interaction1 | A process in which a particle decays or it responds to a force due to the presence of another particle is called interaction. |
| Velocity1 | The ratio of change in position with respect to the time interval over which the change occurred is referred to as velocity. |

177

## Chapter 17. Aerobic Biological Treatment

| | |
|---|---|
| Temperature1 | Temperature refers to measure of hotness of object on a quantitative scale. In gases, proportional to average kinetic energy of molecules. |
| Range1 | The range is the set of values that the dependent variable of a function may take on. A range may be finite as in the set of numbers {1, 2, 3.n} or infinite as in all the mumbers between 0 and 1. |
| Activity1 | Number of decays per second of a radioactive substance is an activity. |
| State1 | Dynamical systems evolve over the course of time. The state of the system at any instant may be identified by the values of certain variables at that instant. For example specifying the angle from the vertical and the velocity of a frictionless pendulum allows us to predict its position and velocity at any future time. Therefore the state of the pendulum at any instant is its position and velocity. In this example the position and velocity are known as state variables. |
| Secondary1 | The output coil of a transformer is referred to as secondary. |
| Function1 | A mathematical function is a rule relating two sets of objects. Here we will restrict ourselves to objects that are numbers or vectors. One of the sets is called the domain of the function, the other is called the range of the function. Functions are frequently expressed as equations as for example Y=X+2. This function is interpreted as follows. For every X in the domain, add 2 to it to get the corresponding Y in the range. Because we are free to choose any X we want, X is called the independent variable. Because once X is chosen Y is fixed, we call Y the dependent variable. |
| Distance1 | Distance refers to separation between two points. A scalar quantity. |
| Liquid1 | Material that has fixed volume but whose shape depends on the container is called liquid. |
| Hydraulic1 | Hydraulic refers to using a fluid as a method of transmitting pressure. It allows forces to be magnified. |
| Particle1 | In 'particle physics', a subatomic object with definite mass and charge . |
| Efficiency1 | Ratio of output work to input work is called efficiency. |
| Origin1 | Origin refers to the point in a reference frame from which measurements are made. It is the location of the zero value for each axis in the frame. |

## Chapter 17. Aerobic Biological Treatment

| | |
|---|---|
| Negative1 | The sign of the electric charge on the electron is negative. |
| Magnitude1 | Magnitude refers to the size of a thing, without regard for its sign or direction. Similar to the absolute value of a number but applies to vectors as well. |
| Slope1 | Ratio of the vertical separation, or rise to the horizontal separation, or run are called the slope. |
| Units1 | The units one uses should be of a size that makes sense for the particular subject at hand. It is easiest to define units in each area of science and then relate them to one another than to go around measuring particle masses in grams or cheese in proton mass units. In particle physics the standard unit is the unit of energy gev. One ev is the amount of energy that an electron gains when it moves through a potential difference of 1 Volt . G stands for Giga, or $10^9$. Thus a gev is a billion electron Volts. The mass-energy of a proton or neutron is approximately 1 gev. |
| Kev1 | One thousand electron volts is referred to as kev. |
| 2+b*x+c . | A function involving the second and lower power, and none higher, of the independent variable. A quadratic function may contain $x^2$ explicitly or it may contain terms like x*, where the second power of x is implied. In general a quadratic may be written as y=a*x |
| Cell1 | In electricity, a cell is a combination of metals and chemicals that produces a voltage and can cause a current. |
| Tracer1 | A small amount of radioactive isotope introduced into a system in order to follow the behavior of some component of that system is a tracer. |
| Phase1 | The particles in a wave, which are in the same state of vibration, i.e., the same position and the same direction of motion are said to be in the same phase. |
| Normal1 | Perpendicular to plane of interest is called normal. |
| Solid1 | State of matter with fixed volume and shape is referred to as solid. |
| Stable1 | Does not decay is stable. |
| Mixture1 | Matter made of unlike parts that have a variable composition and can be separated into their component parts by physical means is a mixture. |

## Chapter 17. Aerobic Biological Treatment

| | |
|---|---|
| Supersaturated1 | Containing more than the normal saturation amount of a solute at a given temperature is supersaturated. |
| Pressure1 | Force per unit area is referred to as the pressure. |
| Gas1 | State of matter that expands to fill container is referred to as gas. |
| Electron1 | Subatomic particle of small mass and negative charge found in every atom is called the electron. |
| Conversion1 | Chemical process turning U308 into UF6 preparatory to enrichment is conversion. |
| Properties1 | Properties refers to qualities or attributes that, taken together, are usually unique to an object; for example, color, texture, and size. |
| Weight1 | Force of gravity of an object is called weight. |
| Gravity1 | A force with infinite range which acts between objects, such as planets, according to their mass is called gravity. |
| Enrichment1 | Physical process of increasing the proportion of U-235 to U-238 is enrichment. |
| Current1 | Current refers to a flow of charge. Measured in amps. |
| Plastic1 | The behaviour of a material is plastic if it does not regain its original shape when the deforming force is removed. |
| Convection1 | Heat transfer by means of motion of fluid is a convection. |
| Force1 | Force refers to agent that results in accelerating or deforming an object. |
| Nominal1 | A nominal dimension is one that gives the intended or approximate size but this may (and often does) vary from the actual dimension. For example, a common lumber shape is a 2x4 but this is a nominal size and the actual dimensions are 1.5" x 3.5". The word is from Latin, of a name, nomin-, nomen name thus can be thought of as 'what we call it'. |
| Random1 | With no set order or pattern, we have random. |
| Power1 | Power refers to rate of doing work; rate of energy conversion. |

## Chapter 17. Aerobic Biological Treatment

| | |
|---|---|
| Heat1 | Heat refers to quantity of energy transferred from one object to another because of a difference in temperature. |
| Density1 | The mass of a given volume of substance. It has units of kg/m3 or g/cm3. When the density is high the particles are closely packed. |
| Speed1 | Speed refers to ratio of distance traveled to time interval. |
| Gradient1 | Gradient refers to the slope of a graph. |
| Friction1 | Force opposing relative motion of two objects are in contact is friction. |
| Turbulence1 | Unstable and disorderly motion, as when a smooth, flowing stream becomes a churning rapid is referred to as turbulence. |
| Kinetic energy1 | Energy of object due to its motion is kinetic energy. |
| Angular velocity1 | The rate of change of angular displacement is called angular velocity. |
| Momentum1 | Product of object's mass and velocity is referred to as momentum. |
| Torque1 | Product of force and the lever arm is torque. |
| Jet1 | The name physicists give to a cluster of particles emerging from a collision or decay event all traveling in roughly the same direction and carrying a significant fraction of the energy in the event. The particles in the jet are chiefly hadrons . |
| Resistance1 | Ratio of potential difference across device to current through it are called the resistance. |
| Event1 | An event occurs when two particles collide or a single particle decay . Particle theories predict the probabilities of various events occurring when many similar collisions or decays are studied. They cannot predict the outcome for a single collision or decay. |
| Rotation1 | Rotation refers to the spin of an object around its central axis. Earth rotates about its axis every 24 hours. A spinning top rotates about its center shaft. |
| Frequency1 | Frequency refers to number of occurrences per unit time. |
| Fluid1 | Fluid refers to material that flows, i.e. liquids, gases, and plasmas. |

## Chapter 17. Aerobic Biological Treatment

| | |
|---|---|
| Set1 | In mathematics a set is a collection of related objects. The mathematical usage is similar to the ordinary English meaning of the word. The objects that make up a set are called the elements of the set. If a set contains an unlimited number of elements it is an infinite set. Otherwise it is a finite set. |
| Movement1 | Change of position is called movement. |
| Position1 | Separation between object and a reference point is a position. |
| Amp1 | The unit of electric current is referred to as amp. |
| Hypothesis1 | A tentative explanation of a phenomenon that is compatible with the data and provides a framework for understanding and describing that phenomenon is a hypothesis. |
| Sun1 | Sun refers to the star at the centre of the solar system. The star the Earth orbits. |

## Chapter 18. Anaerobic Wastewater Treatment

| | |
|---|---|
| Light1 | Electromagnetic radiation with wavelengths between 400 and 700 nm that is visible is called light. |
| Weight1 | Force of gravity of an object is called weight. |
| Heat1 | Heat refers to quantity of energy transferred from one object to another because of a difference in temperature. |
| Energy1 | Energy refers to non-material property capable of causing changes in matter. |
| Temperature1 | Temperature refers to measure of hotness of object on a quantitative scale. In gases, proportional to average kinetic energy of molecules. |
| Efficiency1 | Ratio of output work to input work is called efficiency. |
| Waste1 | High-level waste is highly radioactive material arising from nuclear fission. It is recovered from reprocessing spent fuel, though some countries regard spent fuel itself as HLW and plan to dispose of it in that form. It requires very careful handling, storage and disposal. |
| Gas1 | State of matter that expands to fill container is referred to as gas. |
| Matter1 | We call the commonly observed particles such as protons, neutrons and electrons matter particles, and their antiparticles, antimatter. |
| Solids1 | A phase of matter with molecules that remain close to fixed equilibrium positions due to strong interactions between the molecules, resulting in the characteristic definite shape and definite volume are solids. |
| Electron1 | Subatomic particle of small mass and negative charge found in every atom is called the electron. |
| Conversion1 | Chemical process turning $U_3O_8$ into $UF_6$ preparatory to enrichment is conversion. |
| Gases1 | A phase of matter composed of molecules that are relatively far apart moving freely in a constant, random motion and have weak cohesive forces acting between them, resulting in the characteristic indefinite shape and indefinite volume of a gas are called the gases. |

## Chapter 18. Anaerobic Wastewater Treatment

| | |
|---|---|
| Range1 | The range is the set of values that the dependent variable of a function may take on. A range may be finite as in the set of numbers {1, 2, 3.n} or infinite as in all the mumbers between 0 and 1. |
| Phase1 | The particles in a wave, which are in the same state of vibration, i.e., the same position and the same direction of motion are said to be in the same phase. |
| Pressure1 | Force per unit area is referred to as the pressure. |
| Equilibrium1 | Equilibrium refers to condition in which net force is equal to zero. Condition in which net torque on object is zero. |
| Liquid1 | Material that has fixed volume but whose shape depends on the container is called liquid. |
| Dynamics1 | Study of motion of particles acted on by forces is called dynamics. |
| Activity1 | Number of decays per second of a radioactive substance is an activity. |
| Stress1 | Stress is defined as force per unit area. This is one of the most basic engineering quantities. |
| Stable1 | Does not decay is stable. |
| System1 | Defined collection of objects is called a system. |
| Ion1 | Ion refers to a positively or negatively charged particle formed when an atom or group of atoms loses or gains electrons. |
| Normal1 | Perpendicular to plane of interest is called normal. |
| Quantity1 | A numerical value either scalar or vector, which describes some attribute of an object like its position or its velocity . We sometimes speak of physical quantities to signify that we are talking about an object's properties or attributes as opposed to a purely mathematical quantity. |
| Solid1 | State of matter with fixed volume and shape is referred to as solid. |
| Particle1 | In 'particle physics', a subatomic object with definite mass and charge . |
| Origin1 | Origin refers to the point in a reference frame from which measurements are made. It is the location of the zero value for each axis in the frame. |

## Chapter 18. Anaerobic Wastewater Treatment

| | |
|---|---|
| Hydraulic1 | Hydraulic refers to using a fluid as a method of transmitting pressure. It allows forces to be magnified. |
| Equation1 | An equation is a mathematical expression with an equal sign in it. It signifies that the numerical or vector value on one side of the = is the same as the numerical or vector value on the other side. An equation may include variables and parameters. If any of the variables are rates of change, the equation is called a differential equation. |
| Units1 | The units one uses should be of a size that makes sense for the particular subject at hand. It is easiest to define units in each area of science and then relate them to one another than to go around measuring particle masses in grams or cheese in proton mass units. In particle physics the standard unit is the unit of energy gev. One ev is the amount of energy that an electron gains when it moves through a potential difference of 1 Volt . G stands for Giga, or $10^9$. Thus a gev is a billion electron Volts. The mass-energy of a proton or neutron is approximately 1 gev. |
| Density1 | The mass of a given volume of substance. It has units of kg/m3 or g/cm3. When the density is high the particles are closely packed. |
| Mixture1 | Matter made of unlike parts that have a variable composition and can be separated into their component parts by physical means is a mixture. |
| Period1 | Period refers to time needed to repeat one complete cycle of motion. |
| Vapor1 | The gaseous state of a substance that is normally in the liquid state is vapor. |
| Mole1 | The amount of substance containing $6 \times 10^{23}$ particles is the mole. |
| Decay1 | Any process in which a particle disappears and in its place two or more different particles appear is decay. |
| Velocity1 | The ratio of change in position with respect to the time interval over which the change occurred is referred to as velocity. |
| Second1 | Second is a SI unit of time. |

## Chapter 18. Anaerobic Wastewater Treatment

| | |
|---|---|
| State1 | Dynamical systems evolve over the course of time. The state of the system at any instant may be identified by the values of certain variables at that instant. For example specifying the angle from the vertical and the velocity of a frictionless pendulum allows us to predict its position and velocity at any future time. Therefore the state of the pendulum at any instant is its position and velocity. In this example the position and velocity are known as state variables. |
| 2+b*x+c . | A function involving the second and lower power, and none higher, of the independent variable. A quadratic function may contain $x^2$ explicitly or it may contain terms like x*, where the second power of x is implied. In general a quadratic may be written as y=a*x |
| Set1 | In mathematics a set is a collection of related objects. The mathematical usage is similar to the ordinary English meaning of the word. The objects that make up a set are called the elements of the set. If a set contains an unlimited number of elements it is an infinite set. Otherwise it is a finite set. |
| Unstable1 | Matter that is capable of undergoing spontaneous change, as in a radioactive nuclide or an excited nuclear system. An unstable particle is any elementary particle that spontaneously decays into other particles. |
| Power1 | Power refers to rate of doing work; rate of energy conversion. |
| Sound1 | A longitudinal wave, usually through the air but can travel through liquids and solids is called sound. |
| Event1 | An event occurs when two particles collide or a single particle decay . Particle theories predict the probabilities of various events occurring when many similar collisions or decays are studied. They cannot predict the outcome for a single collision or decay. |
| Gradient1 | Gradient refers to the slope of a graph. |
| Gravity1 | A force with infinite range which acts between objects, such as planets,according to their mass is called gravity. |
| Vacuum1 | Vacuum refers to a region of space containing no matter. In practice, a region of gas at very low pressure. |
| Cycle1 | In wave motion, one cycle is a trough and a crest for a transverse wave, or a compression and a rarefaction for a longitudinal wave. |

## Chapter 18. Anaerobic Wastewater Treatment

| | |
|---|---|
| Core1 | In electromagnetism, the material inside the coils of a transformer or electromagnet is a core. |
| Intensity1 | The amount of radiation, for example, the number of photons arriving in a given time, is called intensity. |
| Fluid1 | Fluid refers to material that flows, i.e. liquids, gases, and plasmas. |
| Plastic1 | The behaviour of a material is plastic if it does not regain its original shape when the deforming force is removed. |
| Random1 | With no set order or pattern, we have random. |
| Function1 | A mathematical function is a rule relating two sets of objects. Here we will restrict ourselves to objects that are numbers or vectors. One of the sets is called the domain of the function, the other is called the range of the function. Functions are frequently expressed as equations as for example Y=X+2. This function is interpreted as follows. For every X in the domain, add 2 to it to get the corresponding Y in the range. Because we are free to choose any X we want, X is called the independent variable. Because once X is chosen Y is fixed, we call Y the dependent variable. |
| Resistance1 | Ratio of potential difference across device to current through it are called the resistance. |
| Movement1 | Change of position is called movement. |
| Nominal1 | A nominal dimension is one that gives the intended or approximate size but this may (and often does) vary from the actual dimension. For example, a common lumber shape is a 2x4 but this is a nominal size and the actual dimensions are 1.5" x 3.5". The word is from Latin, of a name, nomin-, nomen name thus can be thought of as 'what we call it'. |
| Distance1 | Distance refers to separation between two points. A scalar quantity. |
| Secondary1 | The output coil of a transformer is referred to as secondary. |
| Renewable1 | Renewable refers to an energy source that can be used again and will not run out. |
| Tracer1 | A small amount of radioactive isotope introduced into a system in order to follow the behavior of some component of that system is a tracer. |

## Chapter 19. Treatment in Ponds, Land Systems, and Wetlands

| Waste1 | High-level waste is highly radioactive material arising from nuclear fission. It is recovered from reprocessing spent fuel, though some countries regard spent fuel itself as HLW and plan to dispose of it in that form. It requires very careful handling, storage and disposal. |
|---|---|
| Matter1 | We call the commonly observed particles such as protons, neutrons and electrons matter particles, and their antiparticles, antimatter. |
| Decay1 | Any process in which a particle disappears and in its place two or more different particles appear is decay. |
| Conversion1 | Chemical process turning U308 into UF6 preparatory to enrichment is conversion. |
| Weight1 | Force of gravity of an object is called weight. |
| Temperature1 | Temperature refers to measure of hotness of object on a quantitative scale. In gases, proportional to average kinetic energy of molecules. |
| Light1 | Electromagnetic radiation with wavelengths between 400 and 700 nm that is visible is called light. |
| Activity1 | Number of decays per second of a radioactive substance is an activity. |
| Radiation1 | Electromagnetic waves that carry energy are referred to as radiation. |
| Attenuation1 | The process by which a compound is reduced in concentration over time, through adsorption, degradation, dilution, and/or transformation. Radiologically, it is the reduction of the intensity of radiation upon passage through a medium. The attenuation is caused by absorption and scattering. |
| Wavelength1 | Distance between corresponding points on two successive waves is referred to as wavelength. |
| Solids1 | A phase of matter with molecules that remain close to fixed equilibrium positions due to strong interactions between the molecules, resulting in the characteristic definite shape and definite volume are solids. |
| Gases1 | A phase of matter composed of molecules that are relatively far apart moving freely in a constant, random motion and have weak cohesive forces acting between them, resulting in the characteristic indefinite shape and indefinite volume of a gas are called the gases. |

## Chapter 19. Treatment in Ponds, Land Systems, and Wetlands

| | |
|---|---|
| System1 | Defined collection of objects is called a system. |
| Gravity1 | A force with infinite range which acts between objects, such as planets,according to their mass is called gravity. |
| Range1 | The range is the set of values that the dependent variable of a function may take on. A range may be finite as in the set of numbers {1, 2, 3.n} or infinite as in all the mumbers between 0 and 1. |
| Equation1 | An equation is a mathematical expression with an equal sign in it. It signifies that the numerical or vector value on one side of the = is the same as the numerical or vector value on the other side. An equation may include variables and parameters. If any of the variables are rates of change, the equation is called a differential equation. |
| Cell1 | In electricity, a cell is a combination of metals and chemicals that produces a voltage and can cause a current. |
| Retardation1 | Negative acceleration is called retardation. In retardation the velocity of a body decreases with time. |
| Gas1 | State of matter that expands to fill container is referred to as gas. |
| Efficiency1 | Ratio of output work to input work is called efficiency. |
| Hydraulic1 | Hydraulic refers to using a fluid as a method of transmitting pressure. It allows forces to be magnified. |
| Frequency1 | Frequency refers to number of occurrences per unit time. |
| Period1 | Period refers to time needed to repeat one complete cycle of motion. |
| Secondary1 | The output coil of a transformer is referred to as secondary. |

## Chapter 19. Treatment in Ponds, Land Systems, and Wetlands

| | |
|---|---|
| Units1 | The units one uses should be of a size that makes sense for the particular subject at hand. It is easiest to define units in each area of science and then relate them to one another than to go around measuring particle masses in grams or cheese in proton mass units. In particle physics the standard unit is the unit of energy gev. One ev is the amount of energy that an electron gains when it moves through a potential difference of 1 Volt . G stands for Giga, or $10^9$. Thus a gev is a billion electron Volts. The mass-energy of a proton or neutron is approximately 1 gev. |
| Dispersion1 | The splitting of light into its constituent colours is called dispersion. |
| Tracer1 | A small amount of radioactive isotope introduced into a system in order to follow the behavior of some component of that system is a tracer. |
| Distance1 | Distance refers to separation between two points. A scalar quantity. |
| Set1 | In mathematics a set is a collection of related objects. The mathematical usage is similar to the ordinary English meaning of the word. The objects that make up a set are called the elements of the set. If a set contains an unlimited number of elements it is an infinite set. Otherwise it is a finite set. |
| Second1 | Second is a SI unit of time. |
| Dynamics1 | Study of motion of particles acted on by forces is called dynamics. |
| Metal1 | Matter having the physical properties of conductivity, malleability, ductility, and luster is referred to as metal. |
| Liquid1 | Material that has fixed volume but whose shape depends on the container is called liquid. |
| Heat1 | Heat refers to quantity of energy transferred from one object to another because of a difference in temperature. |
| Normal1 | Perpendicular to plane of interest is called normal. |
| Energy1 | Energy refers to non-material property capable of causing changes in matter. |
| Evaporation1 | Change from liquid to vapor state are called the evaporation. |
| Convection1 | Heat transfer by means of motion of fluid is a convection. |

| | |
|---|---|
| Velocity1 | The ratio of change in position with respect to the time interval over which the change occurred is referred to as velocity. |
| Vapor1 | The gaseous state of a substance that is normally in the liquid state is vapor. |
| Pressure1 | Force per unit area is referred to as the pressure. |
| State1 | Dynamical systems evolve over the course of time. The state of the system at any instant may be identified by the values of certain variables at that instant. For example specifying the angle from the vertical and the velocity of a frictionless pendulum allows us to predict its position and velocity at any future time. Therefore the state of the pendulum at any instant is its position and velocity. In this example the position and velocity are known as state variables. |
| Function1 | A mathematical function is a rule relating two sets of objects. Here we will restrict ourselves to objects that are numbers or vectors. One of the sets is called the domain of the function, the other is called the range of the function. Functions are frequently expressed as equations as for example Y=X+2. This function is interpreted as follows. For every X in the domain, add 2 to it to get the corresponding Y in the range. Because we are free to choose any X we want, X is called the independent variable. Because once X is chosen Y is fixed, we call Y the dependent variable. |
| Intensity1 | The amount of radiation, for example, the number of photons arriving in a given time, is called intensity. |
| Power1 | Power refers to rate of doing work; rate of energy conversion. |
| Groundwater1 | Groundwater refers to water found in the voids or free space of soils and rocks underground. |
| Permeability1 | The ability to transmit fluids through openings, small passageways, or gaps is called permeability. |
| Slope1 | Ratio of the vertical separation, or rise to the horizontal separation, or run are called the slope. |
| Properties1 | Properties refers to qualities or attributes that, taken together, are usually unique to an object; for example, color, texture, and size. |
| Shield1 | A mass of attenuating material used to prevent or reduce the passage of radiation or particles is called shield. |

## Chapter 19. Treatment in Ponds, Land Systems, and Wetlands

| | |
|---|---|
| Jet1 | The name physicists give to a cluster of particles emerging from a collision or decay event all traveling in roughly the same direction and carrying a significant fraction of the energy in the event. The particles in the jet are chiefly hadrons . |
| Proportion1 | Proportion refers to two quantities are directly proportional if doubling one of them has the effect of doubling the other. On a graph we get a straight line through the origin. |
| Cycle1 | In wave motion, one cycle is a trough and a crest for a transverse wave, or a compression and a rarefaction for a longitudinal wave. |
| Time period1 | The time taken by a wave to travel through a distance equal to its wavelength is called its time period. |
| Mixture1 | Matter made of unlike parts that have a variable composition and can be separated into their component parts by physical means is a mixture. |
| Phase1 | The particles in a wave, which are in the same state of vibration, i.e., the same position and the same direction of motion are said to be in the same phase. |
| Contamination1 | Radioactive material deposited or dispersed in materials or places where it is not wanted is referred to as contamination. |
| Nominal1 | A nominal dimension is one that gives the intended or approximate size but this may (and often does) vary from the actual dimension. For example, a common lumber shape is a 2x4 but this is a nominal size and the actual dimensions are 1.5" x 3.5". The word is from Latin, of a name, nomin-, nomen name thus can be thought of as 'what we call it'. |
| Movement1 | Change of position is called movement. |
| Gradient1 | Gradient refers to the slope of a graph. |

Clam101

| | |
|---|---|
| Quantity1 | A numerical value either scalar or vector, which describes some attribute of an object like its position or its velocity . We sometimes speak of physical quantities to signify that we are talking about an object's properties or attributes as opposed to a purely mathematical quantity. |
| Efficiency1 | Ratio of output work to input work is called efficiency. |
| Solids1 | A phase of matter with molecules that remain close to fixed equilibrium positions due to strong interactions between the molecules, resulting in the characteristic definite shape and definite volume are solids. |
| Density1 | The mass of a given volume of substance. It has units of kg/m3 or g/cm3. When the density is high the particles are closely packed. |
| Viscosity1 | A measure of how easy or otherwise a liquid can be poured. A high viscosity liquid flows very slowly. |
| Weight1 | Force of gravity of an object is called weight. |
| Gravity1 | A force with infinite range which acts between objects, such as planets,according to their mass is called gravity. |
| Matter1 | We call the commonly observed particles such as protons, neutrons and electrons matter particles, and their antiparticles, antimatter. |
| Range1 | The range is the set of values that the dependent variable of a function may take on. A range may be finite as in the set of numbers {1, 2, 3.n} or infinite as in all the mumbers between 0 and 1. |
| Liter1 | A metric system unit of volume, usually used for liquids is referred to as liter. |
| Function1 | A mathematical function is a rule relating two sets of objects. Here we will restrict ourselves to objects that are numbers or vectors. One of the sets is called the domain of the function, the other is called the range of the function. Functions are frequently expressed as equations as for example Y=X+2. This function is interpreted as follows. For every X in the domain, add 2 to it to get the corresponding Y in the range. Because we are free to choose any X we want, X is called the independent variable. Because once X is chosen Y is fixed, we call Y the dependent variable. |
| Fluid1 | Fluid refers to material that flows, i.e. liquids, gases, and plasmas. |

Chapter 20. Sludge Processing and Land Application

## Chapter 20. Sludge Processing and Land Application

| | |
|---|---|
| Velocity1 | The ratio of change in position with respect to the time interval over which the change occurred is referred to as velocity. |
| Gradient1 | Gradient refers to the slope of a graph. |
| Plastic1 | The behaviour of a material is plastic if it does not regain its original shape when the deforming force is removed. |
| Shear stress1 | Shear stress is produced when two plates slide past one another or by one plate sliding past another plate that is not moving. |
| Stress1 | Stress is defined as force per unit area. This is one of the most basic engineering quantities. |
| Friction1 | Force opposing relative motion of two objects are in contact is friction. |
| Equation1 | An equation is a mathematical expression with an equal sign in it. It signifies that the numerical or vector value on one side of the = is the same as the numerical or vector value on the other side. An equation may include variables and parameters. If any of the variables are rates of change, the equation is called a differential equation. |
| Metal1 | Matter having the physical properties of conductivity, malleability, ductility, and luster is referred to as metal. |
| Vacuum1 | Vacuum refers to a region of space containing no matter. In practice, a region of gas at very low pressure. |
| Pressure1 | Force per unit area is referred to as the pressure. |
| Waste1 | High-level waste is highly radioactive material arising from nuclear fission. It is recovered from reprocessing spent fuel, though some countries regard spent fuel itself as HLW and plan to dispose of it in that form. It requires very careful handling, storage and disposal. |
| Secondary1 | The output coil of a transformer is referred to as secondary. |
| Heat1 | Heat refers to quantity of energy transferred from one object to another because of a difference in temperature. |
| Energy1 | Energy refers to non-material property capable of causing changes in matter. |

## Chapter 20. Sludge Processing and Land Application

| | |
|---|---|
| Liquid1 | Material that has fixed volume but whose shape depends on the container is called liquid. |
| Units1 | The units one uses should be of a size that makes sense for the particular subject at hand. It is easiest to define units in each area of science and then relate them to one another than to go around measuring particle masses in grams or cheese in proton mass units. In particle physics the standard unit is the unit of energy gev. One ev is the amount of energy that an electron gains when it moves through a potential difference of 1 Volt . G stands for Giga, or $10^9$. Thus a gev is a billion electron Volts. The mass-energy of a proton or neutron is approximately 1 gev. |
| Properties1 | Properties refers to qualities or attributes that, taken together, are usually unique to an object; for example, color, texture, and size. |
| Solid1 | State of matter with fixed volume and shape is referred to as solid. |
| Resistance1 | Ratio of potential difference across device to current through it are called the resistance. |
| Mole1 | The amount of substance containing $6 \times 10^{23}$ particles is the mole. |
| Decay1 | Any process in which a particle disappears and in its place two or more different particles appear is decay. |
| Stable1 | Does not decay is stable. |
| Period1 | Period refers to time needed to repeat one complete cycle of motion. |
| Normal1 | Perpendicular to plane of interest is called normal. |
| Second1 | Second is a SI unit of time. |
| State1 | Dynamical systems evolve over the course of time. The state of the system at any instant may be identified by the values of certain variables at that instant. For example specifying the angle from the vertical and the velocity of a frictionless pendulum allows us to predict its position and velocity at any future time. Therefore the state of the pendulum at any instant is its position and velocity. In this example the position and velocity are known as state variables. |
| Hydraulic1 | Hydraulic refers to using a fluid as a method of transmitting pressure. It allows forces to be magnified. |

## Chapter 20. Sludge Processing and Land Application

| | |
|---|---|
| System1 | Defined collection of objects is called a system. |
| Temperature1 | Temperature refers to measure of hotness of object on a quantitative scale. In gases, proportional to average kinetic energy of molecules. |
| Beam1 | Beam refers to a unidirectional or approximately unidirectional flow of electromagnetic radiation or particles. |
| Torque1 | Product of force and the lever arm is torque. |
| Slope1 | Ratio of the vertical separation, or rise to the horizontal separation, or run are called the slope. |
| Gas1 | State of matter that expands to fill container is referred to as gas. |
| Activity1 | Number of decays per second of a radioactive substance is an activity. |
| Adhesion1 | Force of attraction between two unlike materials is referred to as adhesion. |
| Nominal1 | A nominal dimension is one that gives the intended or approximate size but this may (and often does) vary from the actual dimension. For example, a common lumber shape is a 2x4 but this is a nominal size and the actual dimensions are 1.5" x 3.5". The word is from Latin, of a name, nomin-, nomen name thus can be thought of as 'what we call it'. |
| Particle1 | In 'particle physics', a subatomic object with definite mass and charge . |
| Bernoulli's equation1 | In an irrotational fluid, the sum of the static pressure, the weight of the fluid per unit mass times the height, and half the density times the velocity squared is constant throughout the fluid is called Bernoulli's equation. |
| Equilibrium1 | Equilibrium refers to condition in which net force is equal to zero. Condition in which net torque on object is zero. |
| Current1 | Current refers to a flow of charge. Measured in amps. |
| Cell1 | In electricity, a cell is a combination of metals and chemicals that produces a voltage and can cause a current. |
| Centrifugal force1 | Centrifugal force refers to an apparent outward force on an object following a circular path that. This force is a consequence of the third law of motion . |

215

## Chapter 20. Sludge Processing and Land Application

| | |
|---|---|
| Force1 | Force refers to agent that results in accelerating or deforming an object. |
| Speed1 | Speed refers to ratio of distance traveled to time interval. |
| Effective mass1 | The mass in a dynamical system that must be included when we treat the moving parts of the system as though they were a particle, using the free body analysis in applying Newton's laws of motion is effective mass. |
| Angular velocity1 | The rate of change of angular displacement is called angular velocity. |
| Distance1 | Distance refers to separation between two points. A scalar quantity. |
| Symmetry1 | Symmetry refers to property that is now charged when operation or reference frame is charged. |
| Spin1 | The name given to the angular momentum carried by a particle. For composite particles the spin is made up from the combination of the spins of the constituents plus the angular momentum of their motion around one-another. For fundamental particles spin is an intrinsic and inherently quantum property, it cannot be understood in terms of motions internal to the object. |
| Movement1 | Change of position is called movement. |
| Displacement1 | Displacement refers to change in position. A vector quantity. |
| Acceleration1 | Change in velocity divided by time interval over which it occurred is an acceleration. |
| Reprocessing1 | Chemical treatment of spent reactor fuel to separate uranium and plutonium from the small quantitiy of fission products, leaving a much-reduced quantity of high-level waste is reprocessing. |
| Dimensional analysis1 | Checking a derived equation by making sure dimensions are the same on both sides is referred to as dimensional analysis. |
| Permeability1 | The ability to transmit fluids through openings, small passageways, or gaps is called permeability. |
| Dose1 | More specifically referred to as 'absorbed dose', this is a measure of the energy deposited within a given mass of a patient. Absorbed dose is quantified by the unit called the 'rad'. |

## Chapter 20. Sludge Processing and Land Application

| | |
|---|---|
| Cycle1 | In wave motion, one cycle is a trough and a crest for a transverse wave, or a compression and a rarefaction for a longitudinal wave. |
| Neutral1 | Object that has no net electric charge is called neutral. |
| Groundwater1 | Groundwater refers to water found in the voids or free space of soils and rocks underground. |
| Contamination1 | Radioactive material deposited or dispersed in materials or places where it is not wanted is referred to as contamination. |
| Vector1 | Any quantity that has both magnitude and direction. Velocity is a vector. |
| Attract1 | To pull together is to attract. |
| Mercury1 | Mercury refers to the innermost planet in the solar system, and a metallic element that is liquid at room temperature. |
| Time period1 | The time taken by a wave to travel through a distance equal to its wavelength is called its time period. |
| Conversion1 | Chemical process turning U308 into UF6 preparatory to enrichment is conversion. |
| Frequency1 | Frequency refers to number of occurrences per unit time. |

## Chapter 21. Effluent Disposal in Natural Waters

| | |
|---|---|
| State1 | Dynamical systems evolve over the course of time. The state of the system at any instant may be identified by the values of certain variables at that instant. For example specifying the angle from the vertical and the velocity of a frictionless pendulum allows us to predict its position and velocity at any future time. Therefore the state of the pendulum at any instant is its position and velocity. In this example the position and velocity are known as state variables. |
| Activity1 | Number of decays per second of a radioactive substance is an activity. |
| Temperature1 | Temperature refers to measure of hotness of object on a quantitative scale. In gases, proportional to average kinetic energy of molecules. |
| Power1 | Power refers to rate of doing work; rate of energy conversion. |
| Solids1 | A phase of matter with molecules that remain close to fixed equilibrium positions due to strong interactions between the molecules, resulting in the characteristic definite shape and definite volume are solids. |
| Gradient1 | Gradient refers to the slope of a graph. |
| Proportion1 | Proportion refers to two quantities are directly proportional if doubling one of them has the effect of doubling the other. On a graph we get a straight line through the origin. |
| Liquid1 | Material that has fixed volume but whose shape depends on the container is called liquid. |
| Spacetime1 | An extension of the concept of space to include an additional dimension perpendicular to the normal axes spanning our familiar three-dimensional space. This additional dimension measures time so that a point in spacetime locates an event. Since full four-dimensional spacetime is difficult to picture, we frequently work in spacetime consisting of one or two spatial dimensions and the time dimension. |
| Waste1 | High-level waste is highly radioactive material arising from nuclear fission. It is recovered from reprocessing spent fuel, though some countries regard spent fuel itself as HLW and plan to dispose of it in that form. It requires very careful handling, storage and disposal. |
| Distance1 | Distance refers to separation between two points. A scalar quantity. |
| Decay1 | Any process in which a particle disappears and in its place two or more different particles appear is decay. |

## Chapter 21. Effluent Disposal in Natural Waters

| | |
|---|---|
| Dispersion1 | The splitting of light into its constituent colours is called dispersion. |
| Tracer1 | A small amount of radioactive isotope introduced into a system in order to follow the behavior of some component of that system is a tracer. |
| Longitudinal1 | A type of wave motion in which the oscillations are parallel to the direction of wave travel is called longitudinal. |
| Density1 | The mass of a given volume of substance. It has units of kg/m3 or g/cm3. When the density is high the particles are closely packed. |
| Equation1 | An equation is a mathematical expression with an equal sign in it. It signifies that the numerical or vector value on one side of the = is the same as the numerical or vector value on the other side. An equation may include variables and parameters. If any of the variables are rates of change, the equation is called a differential equation. |
| Function1 | A mathematical function is a rule relating two sets of objects. Here we will restrict ourselves to objects that are numbers or vectors. One of the sets is called the domain of the function, the other is called the range of the function. Functions are frequently expressed as equations as for example Y=X+2. This function is interpreted as follows. For every X in the domain, add 2 to it to get the corresponding Y in the range. Because we are free to choose any X we want, X is called the independent variable. Because once X is chosen Y is fixed, we call Y the dependent variable. |
| Velocity1 | The ratio of change in position with respect to the time interval over which the change occurred is referred to as velocity. |
| Second1 | Second is a SI unit of time. |
| Element1 | A pure substance that cannot be split up into anything simpler is called an element. |
| Units1 | The units one uses should be of a size that makes sense for the particular subject at hand. It is easiest to define units in each area of science and then relate them to one another than to go around measuring particle masses in grams or cheese in proton mass units. In particle physics the standard unit is the unit of energy gev. One ev is the amount of energy that an electron gains when it moves through a potential difference of 1 Volt . G stands for Giga, or $10^9$. Thus a gev is a billion electron Volts. The mass-energy of a proton or neutron is approximately 1 gev. |

## Chapter 21. Effluent Disposal in Natural Waters

| | |
|---|---|
| Turbulence1 | Unstable and disorderly motion, as when a smooth, flowing stream becomes a churning rapid is referred to as turbulence. |
| Range1 | The range is the set of values that the dependent variable of a function may take on. A range may be finite as in the set of numbers {1, 2, 3.n} or infinite as in all the mumbers between 0 and 1. |
| Conversion1 | Chemical process turning U308 into UF6 preparatory to enrichment is conversion. |
| Quantity1 | A numerical value either scalar or vector, which describes some attribute of an object like its position or its velocity . We sometimes speak of physical quantities to signify that we are talking about an object's properties or attributes as opposed to a purely mathematical quantity. |
| Matter1 | We call the commonly observed particles such as protons, neutrons and electrons matter particles, and their antiparticles, antimatter. |
| Gas1 | State of matter that expands to fill container is referred to as gas. |
| Cycle1 | In wave motion, one cycle is a trough and a crest for a transverse wave, or a compression and a rarefaction for a longitudinal wave. |
| Barrier1 | Radiation-absorbing material, such as lead or concrete, used to reduce radiation exposure. A primary barrier attenuates useful beam to the required degree. A secondary barrier attenuates stray radiation to the required degree. |
| Normal1 | Perpendicular to plane of interest is called normal. |
| Period1 | Period refers to time needed to repeat one complete cycle of motion. |
| System1 | Defined collection of objects is called a system. |
| Time period1 | The time taken by a wave to travel through a distance equal to its wavelength is called its time period. |
| Stress1 | Stress is defined as force per unit area. This is one of the most basic engineering quantities. |
| Crest1 | The point of maximum positive displacement on a transverse wave is called a crest. |

225

Proof1 | A measure of ethanol concentration of an alcoholic beverage; proof is double the concentration by volume; for example, 50 percent by volume is 100 proof.